CAESAR BELLATOR

Patri mei dedicatus, mea patria prima

REPUBLISHED BY DVX PUBLISHING COMPANY, 2024. FIRST PRINTING 2024.

COVER ART BY CENSORED ANON

INTERIOR DESIGN BY RAG & BONES
https://ragandbones.berserkermail.com/welcome

PAPERBACK ISBN-13: 978-1-957583-19-8

CAESAR BELLATOR

LANCE LEGION

TABLE OF CONTENTS

INTRODUCTION

I

"He who can command, he who is a master by 'nature', he who comes on the scene forceful in deed and gesture - what has he to do with contracts?"

— **NIETZSCHE**

FEW MEN HAVE ever been able to shape the history of a nation more profoundly and dramatically than Gaius Julius Caesar. His exploits are the stuff of myth. It would be considered legend if it weren't for the fact that he so thoroughly documented his actions in such a clear and concise way.[1] The skill of his prose, the aura of his majesty, and the divine quality of his ascendency tends to hide the fact that before Caesar was a god[2] he was merely a man. All great men use power and force to sculpt the world to their satisfaction, using the implements of politics and power to do art on the face of nations.

All power comes from the coercive force of organized violence. In a word it's essence is "martial." The defining delineator between the Aristocrat and the Slave is attitude towards life. The former values valor[3] in combat above all things which, in comparison, seem frivolous and pale. The spiritual Aristocrat and values the honor of arms beyond even his life. The latter values the mere possibility of existing and values it above all things. The former will never suffer dishonor in the form of begging for the scraps of a cold and pallid life. Caesar was a great man first because he valued great exploits and honor which served as an engine for his rise. The profession of arms was his next step towards greatness, and it will be yours too.

Before Caesar took his governorship in Gaul, his military experience was both brief and uneventful. It was only until he was forty-one years of age, nearing the crescendo of his power, that he distinguished himself in a military capacity. For those young of heart among you, remember that greatness is cultivated over time through training and study. Therefore, this will be a dedicated work towards studying his campaigns. His operations are masterful examples of military genius that will never cease to be a font of both practical military lessons and manly inspiration. Many men throughout history have looked back and wished they could have been even a humble legionary to serve such a great man in his exploits. Luckily, we are able to peer back and read his conquest in detail and maybe one day we can leverage his teachings when we don our own boots and face our own valorous enemy before attaining the same undying fame!

AVE BELLATOR!

BOOK I

2

ORIENTATIONS

To give context to Caesar's first actions first we need to talk about Rome's territorial and political disposition at that time. Late Republican Rome was in a position where its great arch-nemesis, Carthage, was vanquished. This transition to a period of unrivaled hegemony, and the corresponding excess in riches and safety, made the need for martial valor and virtue in the citizen body to be less obvious. In turn, the zeitgeist shifted from a focus on virtue to a fixation on the gain of personal position and petty egoistical power. Velleius Paterculus says of this time,

> *"For when Rome was freed from the fear of Carthage, and her rival in empire was out of her way, the path of virtue was abandoned for that of corruption, not gradually, but in headlong course. The older discipline was discarded to give a place to the new. The state passed from vigilance to slumber, from the pursuit of arms to the pursuit of pleasure, from activity to idleness."*

The slip in discipline, one of the featured emphases of *Romanitas* had slipped. In the cultural void, there arose both a period of cynical self-interest and a type of spiritual ennui.[4] Here men of station would take office for their personal ends ostensibly exercised on behalf of the people. The Gracchi brothers broke the original "*constitution*" and function of the Roman Republic, then the Cataline conspiracy arose centered on another Populares pretense, then finally the culminating event which legitimized the forceful seizure of state came in the form of Sulla.[5] Caesar was a product of this proverbial circling the drain, where power was becoming increasingly centralized in the hands of a mighty few who cynically used the approbation of the masses to secure power over the legitimate organs of state through the Senate.

It was his time that Sulla had stepped down from power and the "*normal*" functions of the Roman Senate were restored. Yet, like all events which break the dam, the precedent was irreconcilably set for the rise of an all-powerful dictator who would change the Roman constitution forever. Here, Caesar had made moves to secure this type of absolute power through the shrewd formation of a three-way alliance between Pompey Magnus, Crassus, and himself.[6] Pompey offered the popular support and military renown, Crassus was the richest man in the nation, and Caesar was a Patrician with backing from both the Senate and the People. It was under these circumstances that Caesar chose, after his term as Consul, the governate of three provinces which bordered on the Celtic northern frontiers. Here Caesar sought to enrich himself;[7] more importantly Caesar wanted an opportunity to win military fame and public adoration which could eclipse that of Pompey's and maybe outmaneuver both in the trying for sole power.

This kind of maneuvering was precarious because the ramifications of these actions were apparent to all the Triumvirs. The bond between Crassus and Caesar was already annulled since Caesar was massively indebted to him without any payments forthcoming. The sole bond between Caesar and Pompey was a marriage alliance between their two clans. This marriage is lasted throughout Caesar's exploits in Gaul so we will pick up this story when it becomes necessary. These are the circumstances that Caesar finds himself two years into his five-year governorship of Gaul.[8] His tenure in office was uneventful until... the Celts invaded.

3

THE CELTS INVADE

"We imagine that hardness, violence, slavery, peril in the street and in the heart, concealment, Stoicism, temptation, and deviltry of every sort, everything evil, frightful, tyrannical, brutal, and snake-like in man, serves as well for the advancement of the species 'man' as their opposite."

— **NIETZSCHE**

THE CELTIC PEOPLES, or Gauls as the Romans called them, were among the first enemies of the early Republic.[9] Originally occupying most of northern Italy past the lands of the Etruscans, the Roman Republic successfully either expelled them from those lands or assimilated them into the State.[10] To the immediate north stood the indomitable Alps, a place of soaring rock with sheer edges and only small oases of green pastures for cultivation. Here is where the hardest of all the Gauls lived called the Helvetii.[11]

It is often the case that hard environments make a martial and disciplined people.[12] The reason behind this is because the exigencies of life are so much more difficult for the day to day than it is for the inhabitants of lush and rich environments where both the weather, the earth, and the security from violence is comparatively less. The mountains are environment where those that live must be both strong and resilient to the colds and the lack of arable land. A martial quality is imparted by the constant warring between different tribes and confederations constantly encroaching on each other to seize their respective resources. The only way for people who do not live in a naturally challenging environment to become tough and regimented is through the artificial imposition of hardship and discipline from above.[13] The shock of the coming battles between the Romans and the Celts will show how learned and self-enforced discipline is often superior to the hardness imparted by environmental circumstance.

The Helvetii were naturally a war-loving and conquering people who wanted to expand the limits of their domain beyond their modest territory on the Swiss plateau. Here a leader arose by the name of Orgetorix, who was a man rich in warriors, popularity, and physical wealth. He schemed of the intent to become master of all Gallia by first aspiring to the kingship of the Helvetii.[14] Unfortunately for him, his schemes were thwarted by the neighboring tribes within the Helvetian nation who were jealous of his designs and so imprisoned him where he would quickly die in captivity. This tragedy however served as a sacrificial sacrament which prompted the Helvetii to lift their eyes and fix a common mission to conquer the whole of Gaul. They all gathered together, preparing for a year by both sowing crops as wide as possible as well as gathering to their nation allies from the neighboring mountain passes. After a year, the Helvetii Confederation set out on their quest. They burned their villages and homes in the same way Hernan Cortez burned his ships. The order of their hearts was: Victory or Death.

Now to leave their home territory, there were only two viable passes for the Celts to depart through. One was to their immediate West that passed next to a tall mountain and was along a road which wagons single file could scarcely fit through.[15] The other was through the South via the area of Provence which had recently been pacified by the Romans. Along this latter course there was both food and wide birth for the march to progress quickly along into the Gallic planes of modern-day France. It was along this route that the Celts decided to maneuver and promptly located them at odds with Caesar himself who was master of those territories.

4

CAESAR MARCHES THE LEGIONS

"Surprise, Speed, and Violence of Action."

— **MILITARY MAXIM**

UPON HEARING THE news, Caesar quickly rallies his only legion he has on hand and orders the formation of four more legions to be raised. Hearing that the host aimed to go through his province by the most likely route of Geneva, Caesar immediately commands that the citizens of the town destroy the bridge to deny the Gallic host entry into the territory over the Rhone river. Denying the enemy of this necessary infrastructure[16] the Romans bought themselves time to emplace, train, and create further fortifications along the waterfront 19 miles long to prevent infiltration of Roman territory by forward units of the Helvetii. The tribe attempted to negotiate for military access, and when that failed they attempted to ford the river in boats but were repelled each time. They were left no other option but to extricate themselves north into the territory of the Sequanti[17]out into their intended area of operations.

Strategically speaking the migration of the Helvetii did not just pose an immediate threat of looting and rapine by crossing Roman territory, but also posed a strategic threat that if the Helvetii succeeded in conquering their goals that they would be in control of land which was rich in grain[18] and manpower. The successful transit of the Helvetii prompted Caesar to take matters in his own hands by calling together more troops and creating more legions from scratch. He immediately sallied forth from Geneva and engaged the enemy, slaughtering many who were caught by the speed and violence of the Roman legions. This however proved to be a minor detachment of the Helvetian host who had already cross the river Arar. Caesar recounts here,

> *"There is a river, the Arar, which flows through the territory of the Sequani and the Aedui and into the Rhone... The Helvetii were crossing it by building rafts and tying small boats together. Caesar was informed by his scouts that three-quarters of the Helvetian forces had already crossed this river and only one quarter was left on the same side of the Arar as the Roman army; so during the third watch, he set out from his camp with three legions and came to that part of the Helvetii that had not yet crossed. Attacking them by surprise while they were burdened by their baggage, he slaughtered many of them."*

It is not insignificant to notice the precise series of actions which resulted from this intel. You have to remember how scant good information was, and how much more the fog of war predominated on any ancient battlefield. Here Caesar shows the quality of an excellent strategist and leader by choosing to seize the initiative by exploiting a moment of weakness of the Helvetii. The worst commanders often meet their enemy head-on, but to defeat your enemy in this way in detail, Caesar set the condition for further future successes. Understandably, the Helvetii were alarmed already at the loss of a quarter of their host. Yet it was only when Caesar set his engineers to construct a sturdy bridge across the Arar, crossed it, and pursued the remaining enemy forces did the Helvetii finally succumb to the morale shock and try to sue for terms.

5

PERSPICACITY AND PRUDENCE

"There's daggers in men's smiles."

— SHAKESPEARE, MACBETH, II, 3

OFTEN THE MOST trivial events have the direst consequences. Caesar, before embarking to pursue the rest of the Helvetian host, faced a circumstance where a client tribe of Gauls had failed to provide the necessary provisions of food which the Roman legions required. After some investigating, Caesar had discovered that a shadowy faction of unofficial men with massive popular support[19] had successfully undermined the efforts of the average tribal member to provide the correct amount of grain to the stockpiles. At this point Caesar summoned the leaders of the Aedui to the commander's tent to explain the circumstances of this failure. They demurred when in the presence of a large audience, and so upon receiving them and hearing their public explanation Caesar commanded the rest of the Roman officers and Aeduian delegation members away for a candid personal conversation. This diplomatic approach rendered Caesar the outcome he had desired, essentially naming the underminer, Dumnorix, as the offender.

Checking every detail, Caesar decided to investigate and question other members of this tribe further to gather undeniable evidence as far as Dumnorix's hostile actions.[20] Having acquired them, Caesar then summoned the man's brother, Diviciacus, to a private meeting. Diviciacus was a public leader with official power that had brought up his brother to power in the preceding years before the Roman domination. He had known about his brother's subterfuge but kept quiet out of loyalty. Upon being affectionately and diplomatically confronted by Caesar, Diviciacus burst into tears and confessed the involvement pleaded for the life of his brother hoping to not have too great a retribution exacted on his kin. Caesar assented, both out of compassion but mostly out of astute observation that even though this problem was grave he could not exact merciless revenge as the law promulgated because if he had he risked losing the popular support of the Gallic allies who had chosen to side with the Romans against their kin the Helvetii. It is often the case that entire campaigns are decided on subtle and sensitive actions which can either undermine all your best efforts or make any effort destined for success.[21]

6

SETTING THE CONDITIONS FOR VICTORY

THE ROMANS AND the Helvetii engaged in two decisive actions in the favor of Caesar. Having been completely shattered and utterly defeated on the battlefield, the Helvetii sued for peace. Here is where Caesar as strategic governor thought ahead and was able to translate military success into a strategically political success. Caesar tell us,

> *"He ordered the Helvetii, the Tulingi, and the Larobrigi to return to their countries, where they had started their migration. Because they had lost everything they had harvested and had nothing left at home to keep them from starving, he instructed the Allobroges to supply them with grain. He also instructed the Helvetii to rebuild the towns and villages they had burned. His chief reason for doing this was that he did not want the land they had left to remain empty; it was good land for agriculture, and the Germans living across the Rhine would cross from their own territory into that of the Helvetii and thus become neighbors of the Gallic Province and especially of the Allobroges. The Aedui asked to be allowed to settle the Boii, with their reputation for outstanding bravery, in their own territory, and Caesar gave permission. The Aedui gave them land to farm and later*

> *granted them the status of citizens, equal to themselves in law and liberty."*

This is a vital step that cemented the victory of the Romans over the Gallic tribes. Here, Caesar assured two thing which stabilized the frontier of the Roman Republic: first he gave the Helvetii the sustenance they needed to survive so they don't turn to brigandage for survival and thus compromise the security of the area, second, he had ordered the Helvetii to help rebuild the ruined towns and villages to help the local populace recover and stay healthy and contented and thus more pliant subjects of the Roman State. Learn from this example and see that military success is only useful insofar as you are able to make sure it achieves policy aims.[22]

7

CONFRONTING THE NEW THREAT

QUICKLY UPON COMPLETING the victory over the Helvetians, Caesar was confronted with a new problem which had been festering in parallel to the Helvetian campaign. It had been the custom of the Gallic tribes to take on small mercenary bands from the Suebian tribes to help wage war on their behalf. The Germans became known as those who possessed superior martial skills and won fame for their warlike quality. Small exchanges over time turned into a torrent of Germanic migrants who began forcibly taking dominion over the fertile lands of the Celts in Gaul. A king had come to power among them named Ariovistus.[23] A new host of Suebians had made the crossing at the Rhone river with the intent to extend their dominion over the rest of Gaul outside of the Roman domain. Already, the Ariovistus took roughly a third of the territory of the Sequani and diminished a large number of the Gallic tribes to the position of satrapies. Among the different Gallic peoples the resentment and hatred of the Germans already began mounting to such an extent that positive overtures to the Romans to vanquish this mutual enemy came from the lips of the preeminent Gaul, Diviciacus.

In a meeting with Caesar and all the Roman officials, Diviciacus was driven to explain,

> "*Now that Ariovistus had defeated the Gallic forces in a battle that had taken place near Magetobriga, he was ruling cruelly and arrogantly, demanding as hostages the children of all the noblest men, and inflicting on them, as a warning to others, all kinds of exemplary punishments if everything was not done according to his wishes and whims. The man was barbaric, incredibly easy to provoke, and out of control. It was impossible to tolerate his orders any long. Unless there was some relief coming from Caesar and the Roman people, all the Gauls would have to do as the Helvetii had done: emigrate from their homes, seek another homeland and other settlements far away from the Germans, and face their fate, whatever it might turn out to be.*"

As one can see, the circumstances were dire enough to make the Gauls forget one traditional enemy for another. The stage was set for a confrontation between the Romans and the Suebi.

8

MORALE AND RALLYING

CAESAR ALREADY HAD a reputation from his early career for nightly forced marches and an insistence of getting to advantageous ground first. Here, even before sending out a diplomatic mission to Ariovistus, Caesar already moved his legions towards the area of operations for Ariovistus in an effort to confront him before he received further reinforcements from the Suebi[24] nations and to seize the strategic citadel of Vesontio.[25]

It was upon their arrival at Vesontio, where Roman soldiers were able to confer with their Celtic allies, that the Romans came to hear about the martial superiority and fearsomeness of the Suebi under Ariovistus. Many of the men who were not experienced were thoroughly shaken and began making their last wills and testaments. Rates of "sandbagging"[26] increased tremendously which in turn paralyzed the legions. All this came to Caesar's attention especially because he noticed that the hesitancy and fear began to grip even the most experienced officers and Centurions of the elite Tenth Legion.[27] Mustering all his principle officers and including the Centurions, he addressed them thus,

> *"When all this came to Caesar's attention, he convened his war council, requiring centurions of all ranks to attend as well, and fiercely took them to task – first of all they had considered it their business to ask or deliberate on where they were going to be*

led, and for what reason. When he himself had been consul, Caesar said, Ariovistus had very eagerly sought the friendship of the Roman people. Why should anyone come to the conclusion now that he would abandon his obligations so recklessly?... But if, driven by insane rage, he should start a war, what did they actually have to fear? Why had they lost trust in their own bravery or Caesar's competence? Within the memory of our fathers, the Romans had been confronted by this very enemy. The Cimbri and Teutoni had been beaten by Gaius Marius, and it was generally thought that the army had deserved no less praise than the commander. The Roman army had faced another danger more recently in Italy with the slave rebellion, and these slaves had even been helped to some extent by the training and discipline acquired from us.

From this (Caesar went on), it could be judged how beneficial firmness of courage is. The Romans had feared these men for no reason for a long period, while they were virtually unarmed, but thenlater they had defeated them when they were fully armed and had won many victories. Last of all, these Germans were the same soldiers that the Helvetii had often met in battle and usually defeated, not only in their own but German territory – yet these Helvetii themselves had not proved a match for our army."

This is possible the most instructive piece of advice to men whose morale is wavering. Caesar does a few key things: he *lightly* shames them,[28] tells them that confrontation is unlikely anyway, and reinforces the point that they have already succeeded against worse odds in other combat situations.[29] Finally, Caesar concludes his speech by mentioning that if the legions won't support their commander that he will only take the Tenth Legion who he trusted and leave the rest.[30] The legion commanders and leaders were so moved by this speech that they left the commander's tent both determined and confident.[31]

Caesar then marched out all of his legions, who were puffed up with self-confidence and manly courage, and trod out to meet Ariovistus. Negotiations broke down and ultimately a battle was joined. Through the steel discipline and courage of the Romans, the Germans were routed and were sent running back to Germany. In the process the wives and daughters of Ariovistus were taken as slaves or killed in the stampede. Ariovistus himself, like the coward he was, secured his safe passage across the Rhone river. Caesar wrapped up the campaign then and there, sent his legions to winter quarters, then went over the Alps to catch up on his duties as governor arbitrating legal disputes waiting until the next campaigning season for greater glory.

BOOK II

9

THE BELGAE

"As a matter of self-preservation, a man needs good friends or ardent enemies, for the former instruct him and the latter take him to task."

— DIOGENES

AMONG THE BEST and most fearsome of the Gauls stood the Belgae, who were renowned for their war-making abilities and their successes against other preeminent warlike peoples like the Germans. In fact, it had been common knowledge among the Gauls that the Belgae were not Celts, but in fact Germans themselves, who had migrated across the Rhine from ancient antiquity. Finding the land, they discovered it to be rich for cultivation and hunting, and so the Belgae settled the area by force of arms, making themselves the masters of their territory. Furthermore, it was them alone that kept the migrating Germans from entering into Gaul itself, acting as a shield to the Celts, and thus secured themselves a position of political respect from their Celtic neighbors. It was by their valor and competence at arms that the Belgae held sway over a good third of all territory in Gaul and by which they checked the advance of any threat from the East.

It was during the interlude between campaigning seasons that Caesar had placed his legions in winter-quarters among assorted frontier garrisons in Cisalpine Gaul.[32] It was during this period of relative peace that those Gauls who had not been subdued by Caesar chose to appeal to the Belgae in the hope that this hegemonic power would intercede as a coalition leader. It was here that the situation in Gaul truly started to heat up since it became apparent that the Belgae would answer. Yet, what were their motives for taking up the Gauls on their request? Its not immediately apparent to us, we who are reading from the perspective of the Romans and with the benefit of hindsight, that it could have been the case that there would be no Belgae aggression at all since we take it as a matter of historical fact.[33] This however, is not the case, and Caesar himself makes note of their motives after the fact as thus:

> *"These were the causes of the conspiracy: first of all, the Belgae were afraid that after the pacification of all of Gaul, they would be the next target of our army. Second, they were being approached by number of the Gauls, some of whom, though they did not want any more Germanic involvement in Gaul, were no less annoyed that an army of the Roman people should stay through the winter and become established there. There were also those whose sheer fickleness and lack of constancy induced them to aim at a shift of power."*

As above mentioned, there existed three separate reasons which rallied the Gauls and the Belgae to an anti-Roman cause. It was intuition that gifted Caesar the doubt so as to make the requisite preparations for another year of Gallic hostilities, but it is important for us to learn tools which help us understand that which motivates the hearts of men. Primarily, we must understand one thing: all life at its core is Will to Power,[34] or in contemporary academic terms, it always seeks to maximize power. Hardly is there a case in nature where an organism, much less a political organism like man, is sated with the power he has.[35] Therefore, a sage man like Caesar would have pre-empted the probability that a virile people like the Celts would answer iron with iron and never subject themselves to servitude willingly. Thus both Caesar and the Gauls prepared for war. In dark woods there were secret gatherings. Blue painted men exchanged hostages in good faith and swore solemn oaths under the auspices of the gods. It was under this pretext that Gaul mobilized once again.

10

SEIZING BACK THE INITIATIVE

Before, when the Romans fought disparate Gallic tribes in the south, they had been facing a foe who was at numerical parity to the Romans and who were not as proficient in the art of war as the latter's northern kin. Now, however, the balance and composition of forces aligning against the Romans was quickly recomposing and becoming a force more powerful than any Roman had dealt with thus far. Situated in winter quarters, Caesar had put out feelers[36] across Gaul to get a pulse for where the mood of the Celts were. It was by this good design that Caesar was able to catch wind of the Belgae collating politically and militarily in preparation for the coming martial season.[37] Instead of waiting for the gathering storm to reach its full crescendo of power, Caesar decided to act with all haste. Quickly, he mustered the Legions under his command, and in the middle of great snows and long nights, he marched out like a lightning bolt toward the territory of the Belgae.

Arriving at the periphery of the Belgae, Caesar arrived at the settlement of the Remi, who by virtue of this fast approach, were shocked into submission. Caesar relays,

> *"Since he came there unexpectedly and faster than anyone could have believed, the Remi, who among the Belgae live closest to Gaul, sent Iccius and Andecumborius, the most distinguished leaders of their nation, as emissaries to him. They were to say*

> *that the Remi were handing themselves and all their possessions over to the good faith and power of the Roman people, that they did not agree with the policy of the rest of the Belgae, and that they had no part at all in the conspiracy against the Roman people."*

This, of course, was a farce and a form of obfuscating the general fact: the Remi were faced with either annihilation or submission when it came to treating with the Romans, and they had chosen the latter. All of this was due to Caesar having acted with swiftness and aggression, which offered him the fruits of early victory while diminishing the potential power of his coalescing foe. Furthermore, because the Romans were able to peel away some of the Belgae to their side, they learned valuable intelligence that they would have otherwise been incapable of mustering, in addition to added provisions and fighting power.[38]

Given the fruits of his quick and bloodless victory, Caesar was able to command the Gauls to ravage the lands of the Belgae in a bid to keep all their forces from uniting into one massive host. In other words, faced with the prospect of defeating a far larger foe, Caesar was making every effort to keep contingent forces of that host divided occupied trying to fighting raiding parties of Roman-allied Gauls so that the Romans could concentrate all their the forces on each part of the enemy and defeat him in detail.[39]

In the crush of time, the enormous host of the Belgae approached and quickly were bearing down on Caesar's position. Acting quickly, Caesar went to the periphery of the Remi's territory near a town called Bibrax and fastidiously built a Roman fort on the far side of the river near a Roman-made bridge crossing said river. Roman engineering, implacable as ever, shored up the defenses of that territory making it incredibly difficult for the enemy to assail and simple for allies to resupply. In a series of exchanges, Rome's legions were proven the equal of their enemy, killing many of their number while suffering modest casualties. Succeeding in defending both town and fort for a week news trickled down to the enemy camp that Gallic forces were raiding the homelands of many contingents within the amassed host. Unable to maintain discipline to keep the fight in the favor of the Belgae faced with the irresistible thought of their kinfolk being savaged, much of the barbarian army dissipated reducing their force drastically.

The lack of discipline and order in the barbarian army caused, what was initially a withdrawal, to culminate into a route where no man moved with his unit and was indifferent to the security of their fellow warriors. Diverging into many columns, Caesar exploited this shamble to great effect by pursuing the rear guard of every Belgaen column and killing them in their disorder.[40] Using aggression, discipline, and superior strategy, Caesar was able to disperse this merciless horde and cut them down to pieces. In the hands of a less tenacious or gifted commander, such a situation would have been a disaster since the army of the enemy outnumbered the Romans at a staggering five to one disparity. Therefore, is important to remember that when faced against any great numerical odds, there is always room for victory. Of all martial virtues, it was aggression that Caesar was able to leverage as a force multiplier in exchanges with his enemies and we could clearly see how, though his forces were smaller in comparison, he was able to engender a superior combat power. It was in this fashion, marred by the blood and mud of the Gallic winter months, that Caesar's legions started their lofty ascent to dominating all of Gaul.

II

COURAGE CARRIES THE DAY

"Courage is the wind that drives men to distant shores, the key to all treasures, the hammer that forges great empires, the shield without which no culture exists."

— **ERNST JÜNGER**

CAESAR, UPON HAVING taken back the initiative from the Belgaen coalition, was sailing from one victory to another subduing each tribe after coming across their land and defeating them in detail. Each time Caesar, cultivating the reputation of mercy towards any subdued enemy, consolidated his control on many different tribes by simply forgiving their hostilities after renewing their pledge to the Roman people.[41] The Roman offensive would continue penetrating deep into Belgae territory until finally the Gauls coalesced and engaged the Romans in the shock of battle.

Unlike the tactics of the past, the Belgaens exercised more cunning than was usual for their boorish reputation. Caesar, approaching a previously established Roman fort on the Sabis river, had divided his legions to be in marching order which meant that the baggage train bisected two halves of the host on the road. Due to intrigue by apparent Gallic allies that were riding in the company of the Romans, they informed the Nervii and the rest of the belligerent Belgae of the Roman's critical vulnerability and developed a stratagem to overwhelm the first half of the Roman forces before the latter half could successfully reinforce the first.[42] When Caesar and his vanguard arrived at the river bank that was when the barbarian attack was met! Like an unstoppable black tide of wolf pelts and glittering iron, the Celtic army descended on the exhausted and disoriented Roman legions who were just setting their kit down.[43] There the Romans fought with discipline and determination, however, the pendulum of battle was decidedly against their favor. Legionaries fell in formation, spilling blood which intermingled with that of the barbarians, thinning out with no hope for support in sight. On the right wing, where the legion was most beset by the enemy's main effort, the men were flagging and some were starting to slip away from the rear, forgetting themselves. There, in the nick of time, Caesar rode up on his white stallion, grabbed up a spare shield from one of these wavering legionaries, and charged right into the thick of the Celtic horde. Seeing this great display of courage, determination, and leadership which was in-keeping with the highest Roman martial traditions, the men rallied. Though thin in numbers by then, their renewed onslaught more than repelled the enemy.

Finally, the reinforcing legions of the second half of the host arrived, charging into the thick of battle swinging the scales back in favor of the Romans. Furthermore, the counterattack was so potent that the charge not only carried the enemy back on the field of battle, but a further thrust landed the Romans into the enemy camp, annihilating the possibility of further resistance. Here the victory for the Romans was so decisive that it nearly destroyed, in its entirety, the nation of the Nervii who had put almost all their fighting men in the battle and nearly lost all of them too. Caesar, in his mercy and shrewdness, allowed the valiant combatants to live and ultimately secured control of all Gaul with this last battle. This outcome, however, was not a likely one. Had Fortuna been fickler, it would have been that the legions who were routed before reinforcements had arrived and thereby would have caused a larger defeat of the entire host.[44] What was the factor which overcame an inevitable defeat and transmuted it into a decisive victory? Courage. Courage and discipline, both from the men but most importantly in the leader, Caesar. Men always look to their superiors for guidance, inspiration, and discipline. In short, the leader provides the *imperative*. Its not just a trope to say that a weak leader creates weak followers and vice versa, it is an experience-tested psychological truth that transcends the martial world and penetrates all functions of life. As a leader of men, it is not enough to portray the image of an ideal, one must *be* that ideal, because in the fray of life one's true nature is always revealed. Caesar, after a lifetime of discipline and training, was being revealed to the world as a man amongst men.

BOOK III

12

THE LABOR OF SUBJUGATION

Caesar, having defeated all of Gaul and part of Germania, had returned to the Cisalpine and Illyrian region to take care of his magisterial duties which he had been forced to put off during the tumultuous campaigns of the previous years. This, however, encouraged the newly subjected Gauls to grow weary of their subjugation and try their luck again in the cause of liberty. To Caesar's consternation, first to rise were the Venetii[45] who were powerful sea-faring people that controlled the trade of all positions along the English Channel with extensive political connections with other tribes on both sides of the sea. Then, simultaneously, the Aquetanii rose in the south of the country who were puffed up with the confidence of their norther neighbors rising in rebellion as well. Further, the Alpine Gauls also made their presence felt with a couple of localized uprisings which nearly crushed the Roman garrisons along the Alpine road immediately. All three of these factions were defeated in their turn. Only the Venetii were dealt with harshly because of their flagrant treatment of Roman emissaries.[46] All the other factions were given mercy in the usual manner to secure their loyalty. Nothing about these campaigns are as interesting as the previous two, in fact all three rebelling factions were smaller in comparison to the previous coalitions. What is important to note is the necessary work of empire and what it takes to subjugate a people. Moderns tend to not understand the necessity of constant low-level violence required to put down insurgencies and they often confuse the continued presence of rebellion as a signal of defeat. This is not so, counter-insurgency[47] is a discipline which is on-going which requires perseverance and a strong stomach to do what needed to be done. Romans understood that to keep Gaul they had to seduce them with their culture and lifestyle opportunities but also make sure that all enemies were swiftly beaten. In other words, Rome, whether they like it or not. All of this is achieved with a balance of mercy and ruthlessness, navigated a commander's good judgement. Be as a Roman, and conquer as a Roman does, with discipline.

BELIEVE! FIGHT! WIN!

BOOK IV

13

THE SUEBI: A Martial People

"When asked how one should remain a free man, he said: 'By despising death.'"

— **AGIS SON OF ARCHIDAMUS**

A YEAR AFTER having successfully pacified Gaul in three parallel campaigns, trouble was once again stirred up on the border of the Roman Republic. It was in that year that two Germanic tribes had raided into Gaul to relocate away from a greater threat. And who was this disrupter of the Germanic world? The Suebi.

Caesar relays to us a description of these people:

> *"The Suebi are by far the most numerous and warlike of all the German nations. They are said to have a hundred districts, from each they lead a thousand armed men every year outside their borders to engage in military actions. The rest stay home and provide food for themselves and those on campaign. The next year these in turn serve as soldiers while the others stay home. In this way, neither agriculture nor the knowledge and practice of*

> *warfare are interrupted. Their plots of land are not privately owned or separated from one another, and no one is permitted to stay in one place and cultivate it for more than a year. Their diet does not include much grain but consists mostly of milk and meat of herd animals, and they devote themselves a lot to hunting. Hunting, the type of food they eat, their daily exercise, and the freedom of their way of life – from boyhood they are not in the least accustomed to observing duty and discipline, and they do nothing at all against their will – these all nourish their strength and make them men of immense physical size. Moreover, although they live in extremely cold regions, they have cultivated the habit of wearing no clothing but animal skins, and these skins are so small that much of the body is exposed; they also bathe in [cold] rivers."*

The way of life the Suebi lead should call to mind the habits of the Dorians, or more specifically, the Spartans! Both were of a similar origin, Hyperborea, and both were known for their ruthless discipline and their warrior's contempt for material riches. To many, including the ancient Romans like Tacitus[48], these northmen were a focus of admiration. Austere, hardy, free, and wild, the Germans[49] were what many Romans saw their ancestors to be. That is, the Romans perceived the Suebi to be a great people who were uncorrupted and vital. I believe there is a perennial lesson to be taken from this outsiders observation. To build a strong warrior society, it is important to emulate aspects which were mentioned in Caesar's observations above. The warrior ethos is *nurtured* by being in an environment that is harsh and unrelenting with adversity. The constant exposure to hardship, war, and the elements, presents each man with the necessity to conquer or perish. Furthermore, disdaining material property is a further safeguard against the softening of the Will to Power. We see the salubrious effects it had with the Suebi, however, it is also explicitly stated in the Laws of

Lycurgus which governed a parallel warrior society, the Spartans. Here, Lycurgus sought to encourage the Spartan man to seek wealth in valor and virtu, and not in the effeminate desire for material possessions nor the comfort that they confer.[50] It was this very discipline that made the Suebi a fierce and powerful people and the very reason why they arose as a major contender for power in the Roman arena.

As warriors and military men ourselves, the values indicated above should be foundational elements to building a robust community of heroes. Modernity confers many subtle ailments that, in the eyes of many, are seen as blessings and not curses. We are rich, insulated from adversity of many kinds, and worse, we are sickened by the false value of 'mercy' that makes it impossible to be as ruthless as it is required to rear a strong champion. To become stronger, both individually and collectively, one must have a pitiless heart and favor an ethos of "tough love." In matters of training, one must be as merciless and immutable as Nature can be in harsh climes. In matters of discipline, a soldier must be constant and incorruptible. The war God, Mars, demands nothing less and only those men who give themselves completely to his dictates will be favored by him and be bestowed with his terrible gifts.

14

WAR IS ALWAYS AN EXTENSION OF THE POLITICAL

WHEN THE GERMANIC tribes, fleeing as they were from the Suebi, took lands on the other side of the Rhine from the Gauls it was Caesar who decided to act. There, mustering his legions once more, he quickly marched to the area of operations near the Frisian delta. The quickness of the Romans' approach and the fearsome reputation the legions had already established for themselves by defeating Ariovistus startled the Germanic leadership to send envoys to Caesar in a bid to stay his hand. This, however, was not to be because, during the preliminary negotiations, the Germans skirmished with the Roman cavalry screen and caused a number of eminent allies of Rome with official standing to perish.[51] Caesar, his hand being forced, gave battle to the entire Germanic host. In one decisive engagement, he destroyed the entire nation, killing the men and selling the rest into slavery. This wasn't enough, and therefore felt that it was necessary to cross the Rhine into Germania to make a point.

Caesar explains to us:

> *"With the German war over, Caesar decided for several reasons that he had to cross the Rhine. The most compelling of these was that he wished the Germans – who, as he saw, were so easily tempted to come to Gaul – to be made to fear for their own possessions as well when they realized that an army of the Roman people had both the ability and the boldness to cross the Rhine. Additionally, there was the fact that one part of the cavalry of the Usipetes and Tencteri, as we mentioned above, had crossed the Mosa to collect plunder and grain and had therefore not taken part in the battle. After the rout of their people, they had withdrawn across the Rhine to the territory of the Sugambri to demand that they surrender those who had made war on himself and Gaul."*

The reasons to enter Germania were numerous and yet the motivations were never about adding more Germans to the butcher's bill, this was only incidental, instead it was about extending Roman power abroad to consolidate Roman power nearer home. In other words, the aim for military operations are always political, and therefore social-psychological. Creating the mental image that the Romans were more powerful and more competent than any rival is as much a matter of cultivating that perception as it is actualizing that reality. In furtherance of cultivating this perception, Caesar insisted on building a bridge across the Rhine. The reason being, of course, was to emphasize not just the boldness of crossing into German territory but also to demonstrate the Roman's technological know-how to further demonstrate the power of Rome. Once Caesar and the legions successfully built the bridge, crossed the river, and camped for nearly a month, Caesar decided that his original mission set was accomplished and that to stay any longer would be political diminishing returns.[52] From his position in Germania, Caesar struck camp, crossed back into Gaul, burned his bridge behind him and marched to the English channel. 55

15

NEW HORIZONS

SITUATED ACROSS THE *Oceanus Britannicus* lay an island, lush and green, dimly lit, and shrouded in fog and mystery. For the Romans, this was a land completely unknown. The scope and size of that island was completely uncharted. For Romans in the legions, it was a place of blue demons where spirits from the underworld roamed and tormented passers-by. Only the most savage of men inhabited that distant strip of land who, as Caesar perceived, regularly formed the auxiliaries of Gallic tribes that decided to fight against the Romans. Mystery or not, Caesar was determined, after subduing the Germans, to defeat the Gauls of Britain. To conquer this new land, where no Roman general or legion had previously tread, he believed would be the ultimate boon to his popularity and his political ambition.

Having made the decision, before setting out, Caesar put out feelers for information from the Gallic traders who regularly traded with the distant island, however, they all only knew of the immediate harbors and surrounding areas they sometimes traded in; therefore, there was not much help coming from local guides in the preparations needed to make a successful foray into Britain. Undeterred, Caesar and his legions intrepidly made their quest manifest and embarked on their ships across that foggy sea to the shores of an unknown and hostile place. As soon as he came close to the shore, he could see, as the fog parsed, and gave way to cliffs and beach, men painted in blue picketing the horizon, waiting. Resolute as always, Caesar and his men disembarked and at once were met on the shores with fierce resistance. The fighting almost ended in disaster but, just in the nick of time, further reinforcements landed on that amphibious operation and drove the Gallic tribes back. Making all haste, Caesar and his bone-tired men immediately made camp.[53] Safely ashore with his infantry after a hard-fought victory, Caesar received the emissaries of many different tribes that were suing for peace. They had heard that this was not the end of the Roman column and had succumbed to the inevitable decision of subjugation.

These Gallic emissaries, Caesar received well and in *bona fides*, were quick to conspire and turn on their conquerors since it was shortly after their concluded treaty that they heard a tempest had carried away Caesar's mainland reinforcements when they were in the channel. With rallied determination and vigor, a massive host of Britons manifested again from the mist and renewed their attack. The ensuing battle was hard-fought but ended in a Roman victory, even if it was by the slimmest of margins. During this time, the storm-battered ships of Caesar's fleet had been successfully refitted and were once again seaworthy. Having again sued for peace, the Gauls promised hostages for Caesar as soon as he recrossed the sea again. This however, was a shallow oath and only two tribes followed through with this promise in the end. With the last light of that short campaign, the Romans left, not to make another attempt on that island for another hundred years.

Though the campaign was ultimately a failure, the political currency that Caesar gained from having successfully gone to new lands was massive. From his experiences there, he would write about blue painted Gauls and tell his countrymen back home of quaint stories of chariots being used in battle. Britain would continue to seize the Roman imagination well after Caesar's lifetime until finally, under Emperor Claudius, the Romans would find the initiative once again and take that lush misty island for themselves. On a strategic level, the reason why Caesar had wanted to subjugate the Britons was to deny the Gauls, and the rest of the Celts, of a rear-operating area where they could levy men and monetary support. Undermanned and attacking blind into a territory completely unknown, Caesar was forced to turn back and fall short of his stated aims. In general, I think that the use of the foray into Britain was more for personal political expedience than it ever was about the security imperatives of Rome herself. Nonetheless, the horrors the blue-painted men who moved silently between the trees in the night and who beat deep drums in the hallows of the dense forests, would continue to horrify even the most Stoic Romans till the ultimate evacuation of Roman Britain in the late 5th century. Alas, Britain was not Caesar's glory to add to his mantlepiece of triumph. It was for another time, and another Roman.

BOOK V

16

AN ETHNOGRAPHIC NOTE ON BRITANNIA

The British Isles at the time of Julius Caesar's conquest was not populated by the same people of present-day Britain.[54] The islands were completely populated by the local Celts with no Angle or Saxon admixture that we commonly understand the British to have. Indeed, this was obviously before any Roman colonization which in its turn left very little genetic signature from that time. The question follows then, what were these ancient Britons like?

Caesar describes them here:

> *"Of all these peoples, by far the most civilized are those who live in Cantium, which is entirely a maritime region, and they do not differ much from the Gauls in their customs. Most of those who dwell inland do not cultivate grain but live on milk and meat and clothe themselves with skins. And all the Britons tattoo themselves with woad, which produces a blue color, and this makes them look all the more horrible in battle. They wear their hair long and loose, and every part of their bodies but the head*

> *and the upper lip is shaved. Ten or twelve men share wives; brothers as well as fathers and sons are especially likely to hold wives in common. But the children of these parents are considered to belong to whichever house a woman was first taken to as a virgin."*

Already we can deduce that these peoples that inhabited the islands probably formed from a substrate of pre-Aryan peoples due to the prevalence of their matriarchal practices and their deference to the power of the female anima. Aside from this however, we get a distinct understanding of the Britons as a very backward people both in religious practice and in technological innovation. The Britons were known for practicing human sacrifice, which was a faux pas already with the Romans who made an effort to criminalize the activity in any territory they conquered by that time. Additionally, the use of chariots in warfare, which the Britons were famed to do, had long passed its use in vogue by some 300 years.[55] All considered, its not far-fetched to believe that Caesar's reporting of the Britons to be accurate and fair since he also positively remarks on their bravery and allows for the virtues of his enemies to shine through. Many modern pencil-neck academics who have an axe to grind against Caesar always claim that his commentaries were simple propaganda and fail to mention that Caesar was a noble man and averse to cheap tricks of rhetoric and clout-building; not to mention the fact that he had a number of eyewitnesses who would have undermined the thrust of any egregious political propaganda had he truly reported falsely. With all these factors accounted, Briton seems like a place that hadn't changed much from its previous Bronze Age civilizations and that Caesar's incursion into their territory was a rude awakening of Iron power.

17

AMBUSCADE

"There is only one principle of war and that's this. Hit the other fellow, as quick as you can, and as hard as you can, where it hurts him the most, when he ain't looking."

— **SIR WILLIAM SLIM**

Caesar, after having invaded the British Isles for a second time with limited success, decided to return to winter quarters in his proconsular regions of Illyrica and Cisalpine Gaul. It was during this time that the Germans and the Gauls decided to act on their already smoldering resentment of Roman domination. However, it was only when Caesar was away did the Gallo-Germanic coalition think it was the right time to attack since they had perceived that it was Caesar's ability and commandership that led the Romans from miraculous victory to victory, time and time again. Furthermore, it was only during the winter that the Roman legions were divided and quartered in winter billets which were spread throughout the region. As the commanders of the Romans heard of these enemy movements, a fierce debate rang out among them in the commander's tent as to what the disposition of the army should be. Sabinus, who was adamant that the Romans should rouse from their quarters and confront the enemy before he had time to concentrate and strike. Cotta, on the other hand, urged caution and advised that even if the enemy were to attack any one fort among the Romans, the legions would be able to resist throughout the winter and use the weather and terrain against the Barbarian hosts. He even stated that if the Romans were too impetuous and attacked in these adverse conditions that it would favor the enemy who knew the land to their advantage and who were best acclimated to these conditions. Cotta's sagacious advice was shouted down by Sabinus,[56] and so the army decided to take the offensive course of action.

As the enemy had been observing the Romans' every move, they were able to discern the legion's next move and planned accordingly. Without much need for foresight, the Gauls set a trap for the legion and laid in wait for them to traverse a piece of territory which placed the legionaries in a disadvantageous position. Traveling through the night and encumbered by heavy equipment and a heavier baggage train, the Romans were set upon by the Gauls and were engaged from day-break well into the afternoon. During this time, Sabinus was losing his mind like a chicken without a head and Cotta alone stood his ground like a true Roman and rallied his men to fight. Combat was fierce and everyone sustained many wounds and a great many casualties were accrued during this time. Sabinus, around mid-day, sought a truce with the Gallic commander Ambiorix. Combat continued during this parle and Ambiorix prolonged his speech with no point other than to allow for the combatants in his army to finish destroying the Roman army. The sights of the battle were desperate for the Romans as they were set upon on all sides, unable to come to grips with the enemy but constantly being pelted with missiles and the odd opportunistic blow which felled man after man. The situation became so dire that as many were slain in battle, the few who remained fell on their own swords rather than giving themselves to the dishonor of surrender. The *Aquilifer* of the legion even threw his eagle over the walls of the camp before defended the gates of the fort until he too was felled by Gallic swords. Sabinus, the man responsible for this disaster was summarily killed too.

Having so completely destroyed a Roman legion, Ambiorix was elated and filled with confidence. Quickly after having scored this victory, Ambiorix went from tribe to tribe recounting his great success and made every effort to inflame the hearts of his ethnic brothers with tales of heroism and the promise of vengeful liberty. A flame was quickly picking up from the embers of resentment.

As commanders of men, it is important to understand that leadership is as much about bending to council as it is about decisive action. If Sabinus had listened to Cotta's wise advice, the destruction of his legion, as well as the annihilation of his personal honor, would have been spared. Often, the wisest decisions arise from listening to your subordinates first and only then coming to a complete decision that all men can get behind on. Don't get me wrong, there will be times where you will have to retain control over dissenting opinions and lay the law down, but the way Sabinus had shouted down Cotta was unbecoming of an officer and just plain poor leadership. The trial of war is often a game of chance so it may have been the case that Cotta's course of action would have lead to the same disastrous fate, however, fate should have been revealed after taking the best tac with the correct leadership and risk management. Stories like these are replete throughout military history, so it is doubtful this will be the last, but as a leader make sure you strive to be a Cotta in a Sabinus' command with your wits and honor intact.

18

BELIEVE IN THE MISSION

"In order to convince and inspire others to follow and accomplish a mission, a leader must be a true believer in the mission."

— JOCKO WILLINK, EXTREME OWNERSHIP

QUICKLY AFTER AMBIORIX'S success he capitalized on his victory by going from tribe to tribe within Gaul rousing them to rebellion. Among these tribes who rose to the occasion was the Nervii who were that infamously warlike tribe that numbered among the Belgae's coalition and who almost defeated Caesar only two years prior! Yet, this time, all the tribes of Gaul were simply interested in having their freedoms restored instead of outright looking for revenge. So, when the leaders of the Nervii approached the Roman garrison who were in winter quarters among them with their huge army, they offered very favorable terms of being allowed to leave their territory safely with no harm. Quintus Tullius Cicero, Marcus Tullius Cicero's brother, who commanded the legion there, manfully refused such an offer for cowards. To the delight of the assembled Gauls, and their number was a great many, they quickly built a rampart which surrounded the Roman fort which measured roughly fifteen-thousand feet in circumference. Caesar recounts the scene of these Gauls, who were unequipped for trenchworks, resorting to using their swords to cut up the earth and using their bare hands and the cloaks off their backs to haul it in makeshift fashion. Even thusly underprepared, the host was so massive that they threw up these ramparts quickly, almost as if they were unhindered by their circumstances, and set a siege on the Roman encampment.

Already the balance of battle heavily favored the Gauls, yet the Romans, with their iron discipline, held on. The Gauls on the first day of the siege did not attack, and instead sent riders out to gather all the forces of the Nervii to concentrate on this investment. It was Cicero and his legionaries that took advantage of this respite and quickly improved the defenses of their stronghold. They worked through the night, erecting towers, placing "wall spears" along the walls,[57] and further fortified the cover on the palisades with breastworks made of wicker. Throughout this entire ordeal, from the first contact and parley with the Gauls, Cicero was feverishly keeping pace with his men leading from the front. Already an elderly man who had health issues, he did not stop nor allow himself rest. He was constantly giving orders, ensuring work was being done to his expectations,[58] reassuring his junior leaders, and first and foremost bolstering the discipline and morale of his troops by his mere command presence alone.[59] He only stopped when a throng of men confronted him and effectively forced him to rest.

The Romans, bristling with newly built defenses, were prepared for anything, and were steeled in their spiritual discipline. The Nervii, who had amiable relations with the local Roman garrison, including the commander Cicero himself, did not want to come to blows, and so reached out for a more formal parley. Here, Caesar relays to us:

> *"Then those commanders and leaders of the Nervii who had any reason to think that they were on friendly terms with Cicero and therefore felt justified in seeking to speak with him let him know that they wished to meet in a parley. Offered an opportunity, they brought up the same arguments that Ambiorix had presented to Sabinus. All of Gaul, they said, had taken up arms; Germans had crossed the Rhine; the winter quarters of Caesar and his other commanders were being attacked. They also mentioned the death of Sabinus and point to Ambiorix to corroborate their case. They said that Cicero and his men were mistaken if they hoped for any support from those who were in despair over their own situation. The speakers claimed, however, to be well disposed toward Cicero and the Roman people – so much so, in fact, that they only refused to accept the establishment of winter quarters in their territory: this they did not wish to become a regular habit. As far as they were concerned, the Romans were free to depart from their winter camp unharmed and to set off in any direction they chose, with nothing to fear."*[60]

Caesar then records the virtu of Cicero's reaction:

> *"Cicero had only one answer to these arguments: it was not in the custom of the Roman people to accept terms from an enemy who bore arms."*

This was a stoic and hard-nosed response to such grim odds which were juxtaposed with generous terms for surrender. In this recorded lies the true virtù of Cicero.

In this short snippet of events, here we can observe *real* men on the proverbial "edge of a volcano." Legionaries who were in the wilderness, months away from home, besieged in a foreign town, beset by an impenetrable fog of war, and yet it was in *that* circumstance which they are provided with two choices: manly virtue, or a species of surrender. Yet, despite the horrible circumstances and great temptation, Cicero chose to stand true to his mission and stand his ground. Moreover, his men stood behind him too, in good discipline without any sign of mutiny in spite of the long odds in the face of such reckless peril! So, the question becomes: why? Why choose this tac, especially when these Gauls were known for keeping their oaths, when one could exfiltrate safely? The answer is simple: Cicero believed in his mission called Rome and her Will!

It is the mission which provides the bedrock foundation for any leader. Everything flows from affirming this foundation, which holds true whether one considers the quality of discipline, morale, dedication, the pursuit of excellence in soldiery, or any of the countless facets which characterize the function of a warrior band. Cicero, affirming his charge, solidified not just the resistance of his men but consolidated a stronger defensive posture regarding morale. If Cicero had not believed in his mission, all these things would have fallen away. He would have vacillated on what to do, because, instead of being dedicated to a purpose, he would be effeminately calculating his petty self-interest thus weakening his command's resolve. Not only does unbelief make a leader weak, but one's grip on the one's men slackens because no man can believe in their leader when he doesn't even believe in his own purpose! It was Cicero's resolve to fulfill his mission that would go on to carry the day and save his men.[61] This value is a perennial one for warriors. If you want to be an effective commander, you need to be constantly affirming your mission. When you are beyond the wire, it is that mission which will serve as the golden thread which binds your men together and guides them to their destiny.

Remember legionary, death is certain. Not all actions lead to victory in this life but remember: if you should die fulfilling your charge, *that* is the ultimate triumph, for you have conquered death! 73

19

FRIENDSHIP, COMPETITION, AND VALOR

SEVEN DAYS INTO the siege the situation for the Romans started to become untenable. Totally surrounded, with no hope or chance of relief, the garrison's luck turned from bad to worse. Caesar relays to us a picture of the inferno:

> *"On the seventh day of the siege, an extremely powerful wind arose. The enemy began to launch slingshot made of red-hot clay, along with flaming spears, onto the huts of the camp, which were thatched with straw in the Gallic manner. These quickly caught fire, and the force of the wind spread the blazing straw all over the camp. The enemy raise a mighty shout, as if the victory were securely theirs already. They began to bring up towers and protective roofs and to climb the ramparts with ladders. But the courage and determination of the defenders were supreme: even when they were scared by flames on all sides and pressured by an enormous number of missiles, and although they realized that all their equipment – including their personal possessions – was ablaze, no one stepped down from the rampart to leave the fight, and hardly anyone even turned his head. Precisely, because the danger was so serious, everyone fought most fiercely and bravely."*

And yet, by Mars, the Romans resisted the assault! It was by far the most mauling day out of the siege thus far; casualties abounded and mounds of Gauls were piled up high, right at the cusp of places where they met the ramparts of the wall. Enemy siege towers were set alight, and the great inferno of war roared to ever blazing heights. Smoke filled the blue air, wet mud smattered the kit of every warrior and legionary, sweat and blood mixed and in that chaos it seemed that the Gauls had lost their spirit to fight. Seeing this, two centurions, Titus Pullo and Lucius Vorenus saw the opportunity for valor. Caesar recounts:

> *"In that legion there were two very brave men, centurions, who were now approaching the first ranks, Titus Pullo, and Lucius. Varenus. These used to have continual disputes between them which of them should be preferred, and every year used to contend for promotion with the utmost animosity. When the fight was going on most vigorously before the fortifications, Pullo, one of them, says, "Why do you hesitate, Varenus? or what [better] opportunity of signalizing your valor do you seek? This very day shall decide our disputes." When he had uttered these words, he proceeds beyond the fortifications, and rushes on that part of the enemy which appeared the thickest. Nor does Varenus remain within the rampart, but respecting the high opinion of all, follows close after. Then, when an inconsiderable space intervened, Pullo throws his javelin at the enemy, and pierces one of the multitude who was running up, and while the latter was wounded and slain, the enemy cover him with their shields, and all throw their weapons at the other and afford him no opportunity of retreating. The shield of Pullo is pierced and a javelin is fastened in his belt. This circumstance turns aside his scabbard and obstructs his right hand when attempting to draw his sword: the enemy crowds around him when [thus] embarrassed. His rival runs up to him and succors him in this*

> *emergency. Immediately the whole host turn from Pullo to him, supposing the other to be pierced through by the javelin. Varenus rushes on briskly with his sword and carries on the combat hand to hand, and having slain one man, for a short time drove back the rest: while he urges on too eagerly, slipping into a hollow, he fell. To him, in his turn, when surrounded, Pullo brings relief; and both having slain a great number, retreat into the fortifications amid the highest applause. Fortune so dealt with both in this rivalry and conflict, that the one competitor was a succor and a safeguard to the other, nor could it be determined which of the two appeared worthy of being preferred to the other."*

Here we see an age-old example of male friendship and warrior excellence showcased for us all. Communists, especially in the field of historical academia, have a vested interest in corrupting the past in its recounting because they want to undermine the anything that bolsters a Patriot's spirits in addition to sustaining their false narrative about reality. It is important to let those lies die and ascertain the phenomenon's true essence which lies beyond the imagination of the slave class.

The history of brothers-in-arms has extended beyond even the twin riders of Persia and the Sacred Band of Thebes. It's a phenomenon as old as time which characterizes true brotherly love. Part of male friendship is competition, done in good faith, which challenges the other to strive for Arete.[62] This is precisely the relationship these Pullo and Vorenus had and is foundationally why they were able to inspire each other to reach beyond their ability while simultaneously bringing both honor and a concrete benefit to their legion. In command, it is important to encourage sporting competitions such as these in your men. Sanguine rivalries such as these bring out the best out of every warrior, conferring morale and competency effects which are shared unit-wide, along with the very real spiritual benefits which enrich us all and give pride and meaning to life as a man.

20

BREAKING THE ENEMY

"I am not afraid of an army of lions led by a sheep; I am afraid of an army of sheep led by a lion."

— **ALEXANDER THE GREAT**

THE SIEGE OF Cicero's garrison continued for several more days and the situation became more dire every day that passed. The defense stayed strong but was being severely sapped due to increased casualties and the continued pressure from the Gauls. Finally, their position became so untenable that Cicero decided to send out more riders to contact Caesar, even though the likelihood of their exfiltration through the siege lines was unlikely.[63] Many of the dispatch riders were caught, and in gruesome provocation, were tortured to death in front of the Romans. Cicero, however, ingeniously came up with the idea of using one of the loyal Gaulish auxiliaries to infiltrate through the lines of the enemy and make it past into allied territory. This ruse was successful and quickly arrived to Caesar, who then sent the same rider with a response, who then quickly returned to the siege lines and threw his spear tied with the note back into Roman lines. This maverick-like mission did wonders to improve morale within the garrison, but also was very important in informing the leading commanders of the region as to the desperate situation of Cicero's detachment. Rallying the rest of the Roman assets in the region, Caesar quickly marched his host to Cicero in double quick time. The approach of the Romans alarmed the Gauls who, after convening their war council, lifted the siege and marched their far larger host of sixty-thousand men right at the pacing threat.

The balance of power between the Gauls and the Romans had not much improved in the area of operations. Caesar was still outnumbered against a determined enemy who had a superior understanding of the lay of the land. Therefore, Caesar had to turn both his numerical superiority and the geography of the location to his advantage. Caesar tells us:

> *"The following day at dawn, he broke camp, and when he had advanced about four miles, the enormous army of the enemy came into sight across a valley and a stream. It would have been a great risk for a force as tiny as his to fight on unfavorable ground. Moreover, since he knew that Cicero was freed from the siege, he thought he could reduce his speed without needing to worry. He therefore stopped his advance and fortified his camp on the most advantageous site available. This camp was very small to begin with, since Caesar scarcely had seven thousand men, and none of these with heavy baggage, but he further reduced the camp's size as much as possible by narrowing the walkways, deliberately trying to provoke the greatest disdain on the part of the enemy."*

Thus, forced into a circumstance where the Romans were not able to go toe-to-toe with the Gauls in open battle, Caesar chose to make his weakness into a strength. A few days would go by, and the Romans would hardly react to the encroachment of the Gallic army, who would jeer, make moves to undo the walls, and so on, whereby the Romans would make minimal defensive gestures that sufficed only to maintain their position within the fort. Each time, the Gauls became more and more foolhardy, undisciplined, and unheeding of their superior officers. Finally, at a given signal, when the Gaul's' defensive posture was most vulnerable, the Romans sallied out of all four gates and struck down the enemy in a coordinated lightning strike. The savages who had no discipline in the attack or the retrograde, quickly gave route from what originally had started as a reversal. Completely broken, the Romans were able to chase down that massive horde who were fleeing and kill a great many enemies with almost no casualties of their own. Thus, Caesar, with such a small force, was able to overcome such long odds by using the inertia, ill-discipline, and bravado of the enemy against them.

For commanders to master the ability to judge a situation accurately,[64] having the discipline to stay that course, and having the ability to use the enemy's advantage against him, is a three-fold challenge which recurs often in any military enterprise. Caesar was able to master this adversity and never flinched in the face of superior numbers, since numbers alone confer no distinct advantage. At the end of the day Caesar once again shows how one man can be one-million on any battlefield, no matter the odds. It is important to remember that you too can be one-million.

BOOK VI

21

THE WARRIOR ETHOS

"The Commander of your Senses: Discipline, ferocity in the face of temptation, resistance to one's lower self, flogging it to death where necessary, refusing the promises of decadence, antipathy to notions of pleasure. A Man that won't lapse, a Man that is tenacious."

— NIETZSCHE

WHAT ETHOS GIVES rise to the creation of the best warriors? What type of culture gives rise to a martial people? These are questions which have beset commanders and scholars of the military arts for generations. The sum of a man's qualities and discipline can be said to be bred into him over generations. A warrior's breeding is informed by the natural environment of his homeland, the cultural mores of his society, his natural dispositions in body and mind, and the overall rigor of military instruction provided to him. This subject is not a new one, nor is it a long-abandoned notion, and we can trace the origin of this military wisdom to even before the Romans; the origin of our quest can be found deep in the cave of the Pythian seeress, where Lycurgus consecrated his people's destiny.[65]

The best warriors are those that devote themselves to a harsh ethos. A standard of discipline, austerity, privation, adversity, and the stolid embrace of trial and misfortune. Strength, honor, and victory naturally follow from a rigorous devotion to such principles. Furthermore, it is Caesar that affords us the opportunity to see those values' efficacy demonstrated. During his operations in Gaul, Caesar was given a stark juxtaposition between the major nations of Europe which inhabited the region around Gaul at that time. Numbering among them stood two major peoples, the Celts and the Germans. Both were known as warlike and ready to confront men in battle with an aim to win glory and honor to their name. However, the Gauls, who formed the nations of those Celts in Gaul, had over generations lost their potency in war-making. The Celts were the first Indo-European nation to cross into Europe and Asia minor, imposing themselves over the previously native populations, who won their dominion through force of arms. It was through this warlike excellence that the Celts maintained their realm for more than 1,500 years. However, the more of self-discipline, which sustained their martial power, had steadily eroded by succumbing to the vices associated with their success. They slowly eschewed habits of frugality, discipline, fierceness in spirit, and the prizing of triumphs over adversity and replaced them with effeminate habits like seeking material luxuries, slavishness to peace-time society, and consummate selfishness in hopes of petty distinction over the excellence of their nation. By the time the Germans had coalesced and descended into

Europe, the Celts had become a much-weakened people. The Germans themselves, who are a nation created in the cold and adverse conditions of the far north, still practiced a warriorlike askesis, the same original formula since abandoned by the long-reduced Celtic nation.

War is one of those phenomena that distills the nature of life down to its most pure and naked forms. Life is, at base, an arena of competing wills, chaos, and truth revealed by action. In war, as it is in life generally, material circumstances are variable. Supplies are used or lost, technology works under circumstance, the fog of war throws into doubt the caprice of the moment, and men die at the whim of luck just as often as they are preserved by their own valor. In this chaotic environment, warriors sup from the cup of conquest if they can adhere to the maxims of discipline, regardless of the circumstances, at every moment. Lesser beings - who value only the material and this life - ultimately become slaves since their lives are dictated entirely by their environment. They abdicate their souls and measure their fulfillment in the accumulation of goods or by the longevity of their existence. Therein lies the fatal mistake: by valuing a good outside one's soul, one has compromised their position and made themselves vulnerable to being manipulated. From there, cowardice rears its ugly head and poisons the soul with lethargy, petty jealousy, avarice, weakness, and cumulates into abject servility. This road downward is slow yet precipitous, nonetheless. It was by this path that the Celts allowed themselves to be seduced, and by which they were induced to cede their freedom to the Germans despite their late desperate efforts. Caesar corroborates this fact with his own observations saying, "As for the Gauls, however, the nearness of our provinces and the familiarity they have developed with imports from overseas have bestowed on them many things to make their lives more agreeable and lavish. They have gradually become accustomed to losing in war and, having been beaten in many battles, no longer even compare their courage with that of the Germans."

As indicated previously, the Germans had long become the superior of the Gauls. Their formula for success, as Caesar remarks subsequently, shows us in no uncertain terms why that is:

> *"The German way of life is very different. They have no druids to preside over matters related to the divine, and they do not have enthusiasm for sacrifices. They count as gods only those phenomena that they can perceive and by whose power they are plainly helped, the Sun, Fire, and Moon; others they do not know even from hearsay. Their whole life is spent on hunting and military pursuits. From early childhood they devote themselves to toil and endurance. The men who abstain the longest from sexual activity enjoy the greatest praise among their own people. They think that abstinence enables men to grow taller and that it enhances their strength and muscles. They count it among the most shameful things to have experience of a woman before the age of twenty: but there is no prudery or ignorance in sexual matters, because the two sexes bathe together in rivers and wear animal skins or small pieces of reindeer hide, which leave most of their bodies bare."*

All these enforced habits of the Germans were imposed so it could imbibe key habits in its people. Their environments were designed to emplace artificial hardships, like a gym for the soul, to train them to have contempt for pain, pleasure, and even death itself. Strengthened against these illusions, the Germans set a foundation upon which they built a powerful war machine. In place of wealth, these wolfmen esteemed valor in arms and the acclaim of their peers as the greatest good in life. Their bounty was held in the confines of their soul, as were the means to its accumulation. This ethos, therefore, strengthened their willpower and encouraged self-mastery. Being masters of their own souls, they quickly manifested themselves as the masters of others.

We warriors and commanders of men would do well to bear in mind that this formula is true regardless of time or place. As I stated before, it was by a similar lifestyle that the Spartans under Lycurgus were able to transform themselves from a weak polis into one of the greatest warrior societies the world has ever seen. These mores are what made the Prussians not only great militarily but also exceptional in the field of technological advancement.[66] All the great nations of the world held these principles in common. Romans, Huns, Germans, Mongols, Japanese, and so on, all lay their laurels of victory at the feet of Mars. All fell when they succumbed to the siren song of peace and the ease of plenty.

22

CAESAR, GALULA, AND THE ART OF SUBJUGATION

"Vae Victus"

— **ROMAN PROVERB**

THE DISTANCE OF time and peoples allows the students of history to reflect on the past with a certain emotional lucidity. Of course, there will always be an element of bias when reflecting on one's own history or those causes that reflect the interests of one's contemporary interests. When analyzing the strategy of war, it is important that we maintain a spiritual discipline which allows us to analyze any phenomenon with cold indifference when called for. With all of this in mind, the question of conquest and subjugation of peoples often evokes a sense of womanish moral indignation. We warriors must not fall for this type of emotionality.

Before we continue I will address the emotional zeitgeist of "Just War" directly: War is inherently brutal, austere, and callously pragmatic. War is only limited by the political ends a combatant is aiming to achieve. Regulations and "Laws of War" are ultimately conceits that are a shadow of the *calculus of means*. Mercy and ruthlessness are simply tools that are deployed when applicable to attain a certain goal. A military commander needs to be ready to employ both, unemotionally, when the situation calls for it. Neither saints nor sadists are apt practitioners of war and are not welcome in this discourse. Full stop. The only first principle a warrior knows is thus: "do good to my friends, destroy my enemies."

Returning to the matter at hand, when militaries have succeeded in destroying the enemy's conventional means of force, there follows a period of imposing a new political order over a nation's new subjects. In modern parlance, this phase of military operations is called "Occupation". A term more honest to the meaning of this phenomenon is "Subjugation" which defined means, "To make subordinate or subject to the dominion of something else or one's power." Occupation is a sanitized term crafted to be more palatable to the ears of the peasant and war-averse lower castes, but luckily we can dispense with that. Therefore, when political power has been established and created a new status quo, any political upheaval is thus called a "Revolution." Galula explains:

> *"A revolution is an explosive upheaval – sudden, brief, spontaneous, unplanned (France 1789; China 1911; Russia 1917; Hungary 1956). It is an accident, which can be explained afterward but not predicted other than to note the existence of a revolutionary situation. How and exactly when the explosion will occur cannot be forecast. A revolutionary situation exists*

today in Iran. Who can tell what will happen, whether there will be an explosion, and if so, how and when it will erupt?"[67]

In the context of Caesar, a Roman commander of forces within a newly acquired territory, it may seem to our ears like a case of simple revolt against a foreign force, and you may be right in an emotional sense, however it is only within the context of Galula's revolution that we are able to diagnose the methods of successful subjugation Caesar will later lay bare for us.

To understand the point of war, especially a war of revolution, it is important to understand what the strategic goal being fought over is the submission of a people. In a conventional war, the core element's primacy, the nation, is not apparent because the level at which one party or the other will concede defeat is layers away from the central axis of the political origin. Therefore, the central element is hardly ever the focus of a conventional campaign. However, in the case of conquest, the outer layers of political capital have been stripped away and thus we are left with fighting over terrain which is the most fundamental political ground: the people. Galula goes further to explain the dynamic:

> *"Afflicted with his congenital weakness, the insurgent would be foolish if he mustered whatever forces were available to him and attacked his opponent in a conventional fashion, taking as his objective the destruction of the enemy's forces and the conquest of the territory. Logic forces him instead to carry the fight to a different ground where he has a better chance to balance the physical odds against him.*
> *The population represents this new ground. If the insurgent manages to dissociate the population from the counterinsurgent, to control it physically, to get its active support, he will win the war because, in the final analysis, the exercise of political power*

> *depends on the tacit or explicit agreement of the population or, at worst, on its submissiveness.*
> *Thus the battle for the population is a major characteristic of the revolutionary war."*[68]

It is contending on the grounds of winning the people of Gaul that Caesar plans his numerous campaigns around. The multiple uprisings of the Gauls, the interventions by the Germans and Britons, are all challenged by the Romans who strike at control of the people and their political elites as much as they do fight the conventional forces of these various nations on the battlefield.

In sum, one must fight for the people and capture them politically, either through justice and reward or by ruthless annihilation, if one seeks to conquer a nation.

23

FORTUNE FAVORS THE AUDACIOUS

It was shortly after Caesar had driven out the Germans that a certain Gallic nobleman by the name of Ambiorix was agitating for rebellion. As it has been stated before, the temperament of the Gauls was always jealous of their freedoms, quick to rise to rebellion, but reluctant to persevere through adverse conditions should victory not immediately follow the aims of their war efforts. Caesar, who by this time was a time-tested experienced commander in the Celtic theater, knew that to prevent this uprising before it spread, he had to make all haste to intercept Ambiorix and his retinue before the rebellion grew in manifold numbers. Caesar however, was detained by having to wait for the wheat to ripen and therefore could only send his advanced guard of cavalry, his only asset not logistically arrested, to intercept Ambiorix at Atuatuca. Caesar thus gave the order to his subordinate officer, Lucius Minucius Basilus. Here Caesar relays to us his plan of attack:

> *"He sent Basilus ahead with the entire cavalry through the Ardennes forest. Caesar hoped that Basilus might achieve success by the speed of his advance and by exploiting opportune moments. He admonished him to prohibit the use of fire in his*

> *camps so that his approach would not be signaled from far away, and said that he would follow as fast as possible."*

The Roman forces closing in on the newly resurgent Gallic threat were triumphant. Quickly making the advance, in keeping with admonishments of his captain, Basilus was so quick in his advance that he caught up with Ambiorix himself. Though the action was indeed a success, so much about the outcomes of war is up to Fortuna herself, and therefore from the clutches of ultimate victory, triumph had been postponed. Caesar explains this chaotic truth reflecting on this action here:

> *"Basilus did as he was ordered. He covered the distance more swiftly than anyone could expect and caught many people by surprise who were still in their fields. Upon receiving information from them, he pressed on to capture Ambiorix himself, in the place where he was said to be staying with only a few horsemen. Fortune plays a great role in everything but has an even greater influence in warfare. For, on the one hand, it is a remarkable piece of chance that Basilus came upon the man himself when he was not expecting it and unprepared, and that Basilus appeared in the sight of all those present before any message or rumor had announced his approach. On the other hand, it was a great instance of good fortune for Ambiorix that, although he had been deprived of all the military equipment he had with him, and the chariots and horses had been seized, he himself should be able to escape death. But this is precisely what happened, because his building was surrounded by woods – as in fact, most Gallic homes are, since the Gauls tend to seek out places near woods or rivers to avoid the heat – and his followers and clients, given the narrow space, were able to withstand the assault of our cavalry for a short time. While they were fighting, one of his men threw him on a horse, and the forest hid his*

> *flight. In these ways, therefore, Fortune proved to be very powerful, both in subjecting him to danger and in allowing him to escape from it."*

Fortune is a goddess which lives on till this very day. For those veterans of the Global War on Terror, they know only too well that the speed of action bore them fruits which resulted in battlefield success, and even then, on some occasions, Fortune intercedes on the side of the enemy, allowing him to slip through the fingers of fate. This is a consideration you must account for in your own plans of battle and an element of battle you should steel your mind against so that sudden changes do not stun you into inaction.

24

IN VICTORY STAY VIGILANT

"Walk my post in a military manner, keeping always on the alert and observing everything that takes place within sight or hearing."

— **USMC, SECOND GENERAL ORDER OF THE SENTRY**

THE ENSUING CAMPAIGN against the Eburones was a resounding success for the Roman forces. Yet, part of that success was due to the involvement of Gallic allied forces, which Caesar had invited to despoil the insurgent enemy due to the tactical considerations which were imposed. One of the negative externalities caused by the chaos of involving neighboring tribes to raid the enemy is the fact that it invites unwanted guests. Caesar informs us thus:

> *"...the news happened to reach the Germans across the Rhine that the land of the Eburones was being devastated and that everyone was actually being invited to take part in plundering their possessions. The Sugambri, who live closest to the Rhine ... assembled two thousand horsemen. They crossed the Rhine with*

> *boats and rafts thirty miles downstream from the spot where the bridge had been built and the garrison left by Caesar. They approached the territory of the Eburones, caught many of the people who were scattered in their flight, and took a large number of herd animals, for which barbarians are very greedy. Lured by the prospect of more spoils, they advanced farther. As they were born warriors and bandits, no swamps or forests could slow them down. They interrogated their captives as to where Caesar was. They found out that he had gone away a considerable distance and that his whole army had left the area. One of the captives said, 'Why are you pursuing this pathetic, scanty loot when it is within your power to become the most fortunate men on earth? You can get to Atuatuca in three hours. The Roman army concentrated all its possessions there. The garrison is so tiny that it cannot even deploy men all around the wall, and none of them dares to go outside of the fortifications.' Once they knew of this opportunity, the Germans left in hiding places the plunder they had taken. They hurried to Atuatuca, guided by the same man who had provided them with this information."*

Quickly making all haste and taking advantage of the terrain features to conceal their advance, the Germans rushed upon the gates of the Roman camp with such surprise that even the merchants who had set up shop just under the ramparts had no time to make their escape. The camp, which was being held by a reserve element of Roman soldiers, was guarded just well enough to resist the attacks after multiple onslaughts.[69] Upon hearing the sudden commotion, a detachment of five cohorts which had been sent out to reconnoiter the area around the camp had rushed back and attacked the Germans in the flank. Fearing being overcome, due to the intrepidity and aggressiveness of the reinforcing cohorts, the German raiders broke contact and retrograded away from the Roman camp.

Much can be said about this little episode, but the most important aspect can be reduced thusly: though it may seem that victory is at hand the enemy is vanquished, is the duty of every soldier to always keep on the alert. Had it been the case that the troops garrisoning the fort were of lesser discipline, the soldiery of the fort would have been overwhelmed and a disaster would have ensued. It is precisely for this reason that discipline is never a question of when it is useful, it something which must be kept and honed at a constant state of readiness regardless of your perceived situational necessity. To be a warrior means to be constantly ready and constantly disciplined.

BOOK VII

25

WEAKNESS BREEDS CONTEMPT

"What belongs to greatness - Who will attain anything great if he does not possess the strength and the will to inflict great suffering? Being able to suffer is the least thing: weak women and even slaves often attain mastery in that. But not to perish of inner distress and uncertainty when one inflicts great suffering and hears the cry of his suffering - that is great, that belongs to greatness."

— NIETZSCHE

WHEN LEADING MEN, indeed when leading any organism, pain is more important than euphoria ever can be. There is a misconception that imposing pain should be avoided; that causing pain lessens both the master and the subordinate in equal measure. Nothing could be further from the truth in terms of inflicting pain in the service of constructive ends, even if they are selfish.[70] However, the biggest mistake of those masters of men who impose pain on their subjects, who then find themselves not equal to their self-perceived crime, who then renege only add only insult to injury to their act.

It may be bad enough to be dominated by another man, it is worse to be the subject of dominion by someone who is not worthy. Just as it characterizes the social issues of today where the Aryan peoples of the earth have become unequal to their reign over the lesser peoples of the world and therefore invites danger against them, so too did the Gauls find the Romans unequal to their dominion over themselves and thus catalyze a great revolt.

Returning to the campaign, shortly after crushing Ambiorix's revolt, Caesar returned to Italy to commit to his civilian responsibilities as governor and adjudicator for the territories of his charge. His newfound political balance achieved in Gaul was rudely swept away by disturbances in Rome. Caesar explains:

> *"With Gaul finally calm, Caesar set out for Italy to preside over the local judicial meetings, as was his established practice. There he learned of the murder of Publius Clodius and the Senate's decree that all young men of Italy should take the oath of military service, and so he, too, determined to levy soldiers throughout the province. News of these developments quickly reached Gaul beyond the Alps. On their own, the Gauls expanded on the rumors they heard and came up with an invented story, which the circumstances seemed to make inevitable: that Caesar was being held up by unrest in Rome*

> *and, because the conflicts there were so great, could not come to join his army. Those who, even before this, had expressed resentment about their subjection to the rule of the Roman people were incited by this opportunity to start to make plans about war more freely and boldly than before. Gallic leaders called meetings to confer with one another in remote places hidden in the woods. They complained about the death of Acco, emphasizing that they too might suffer his fate. They commiserated about the shared misfortune of Gaul. Proposing all kinds of promises and rewards, they were urgently searching for men who were willing, at the risk of their won lives, to unleash a war and take up the cause of restoring liberty to Gaul."*

Whether one is a leader among one's own countrymen, the Dominus of subject nations, or even the commander of one's own troops, weakness always breeds contempt and emboldens those resenting elements within one's dominion to take power.[71] It is precisely this type of contempt which would go on to destabilize Gaul against Caesar and raise one of the most historically powerful rebellions Rome has ever known. Therefore, as a leader, it should be incumbent on you to always portray yourself as strong, better still, it is imperative that you are always strong to the best of your ability.

26

NO BETTER FRIEND, NO WORSE ENEMY

"A prince is also respected when he is either a true friend or an absolute enemy, that is to say, when, without any reservation, he declares himself in favor of one party against the other. This course will always be more advantageous than standing neutral. If two of your powerful neighbors come to blows, you have either to fear the winner or not. In either case it will always be more advantageous for you to support one of them and to actively make war.

If you do not declare yourself, you will invariably be attacked by the conqueror, to the pleasure and satisfaction of the loser, and you will have no reasons to offer, nor anything to protect or to shelter you. The conqueror does not want doubtful friends who will not aid him in the time of difficulty, and the loser will not protect you because you did not willingly, sword in hand, follow his fate."

— **MACHIAVELLI, THE PRINCE**

UPON HEARING OF Vercingetorix's revolt, Caesar made all haste to meet the threat in force to drive the enemy on to his back foot. Taking his host through the snow-banked Alps, Caesar achieved surprise over the Gallic enemy and killed many who were taken unawares and unprepared.[72] Through a series of skirmishes, Caesar took back control of the territory of Province and a few other key redoubts. Vercingetorix's response to such a sudden and violent rush of attacks by Caesar was predictable. The Averni then chose to attack at Caesar's most important allies within Gaul, the Bituriges, who were tribute paying to the Aedui and clients of Rome respectively.

The move was designed to drive a wedge between Rome and her supporters while encouraging the disaffected and liberty loving Gauls to take up arms against her. It is here that Caesar was faced with a dilemma:

> *"This move posed great difficulties for Caesar as he tried to form a plan of action. If he kept his legions in one place for the rest of the winter and the tribute-paying dependents of the Aedui were conquered, this would, he feared, cause all Gaul to revolt, because it would appear that Caesar could not be relied upon to protect his friends. If, on the other hand, he brought the army out of its winter quarters too early, his concern was that he would incur great problems with his food supply, because transport would be severely hampered. Yet it seemed preferable to endure every kind of hardship rather than suffer a huge disgrace that would alienate the goodwill of all his supporters. There he urged the Aedui to deliver the supplies and sent messengers ahead to the Boii to tell them he was coming and to encourage them to remain loyal and hold up bravely against the enemy attack. Leaving two legions at Agedincum with the baggage of the entire army, he set out for the territory of the Boii."*

Faced with a strategic decision between facing the danger of starvation and bad weather or the danger losing support among his allies, Caesar made the right call and forced marched in the defense of his friends. This quandary is an eternal recurrence which faces many princes and commanders alike. A selfish and undynamic military mind considers the "military science" behind preserving logistical and "force ratio" principles of a unit over the political considerations those respective units are trying to accomplish.[73] Considering the unit capabilities and limitations are important but they must be stretched beyond what they are predicted to be if one wants to wage a successful campaign.[74] Ultimately, we have to remember that victory during a war is always doubtful and the outcome is always fluid. You can make as many preparations as possible, be cautious as possible, and be as drilled as possible and still fail. Therefore, in keeping with the Probablistic view of war,[75] we should aim to hold true to our core principles while following through on each opportunity and dilemma dictated by the exigencies of the strategic outcome we desire.

Finally, in this example where we measure the merit of friendship and the strength of our word, it should be known that a moral code of fidelity to friends makes itself apparent here. Fidelity to friends is not always easy, defending them can often lead to our personal ruin, however, life and war is often won by those bands of men who can stick together, and defeat is a chalice the selfish and oath breakers always sup from. Therefore, legionary, be always a steadfast and powerful friend to those you have honored with your word, not simply because victory comes from such actions of virtù but because, even if setbacks occur due to your dedication, you will have scored a victory of excellence[76] over life and made known to both friend and foe alike the quality of your fiber and the value of your leadership!

27

DENY THE ENEMY

"What benefits the enemy, harms you; and what benefits you, harm the enemy."

— MACHIAVELLI

THERE ARE FEW lies in our time which are more pernicious or pervasive than the twin falsehoods, "life is not zero-sum" and "resources are not zero-sum." Neither of these assertions could be further from the truth because both are finite resources and are often in extremely limited supply. Therefore, as a military man, you must part ways with the delusions of eunuchs and economists and reembrace the terrible beauty of reality. Be aware that every asset your enemy holds is one they have over you; likewise, every power you grasp is one that confers advantage over your enemies. It is precisely for this reason that in combat operations one should seek to deny the enemy any and all forms of aid or material advantage so that one may undermine the strength of the forces arrayed against you and bolster one's own.

It was precisely along these lines that the able commander of the Gauls, Vercingetorix, leveraged his position as the local controller of food stuffs to use hunger as a weapon against the oncoming Romans. Caesar, at this time, had been forced to raise two more legions that were stationed in Gaul for winter quarters in his campaign against the insurrectionist Gauls. Mobilizing more troops to the war effort also means something else: the logistical requirements of food to sustain these men also increases. This is why Vercingetorix wanted to turn Caesar's center of gravity against him since the Gauls would have provisions of their own and Caesar would be forced to burn through his scant supply at a now accelerated rate. Caesar relays to us Vercingetorix's stratagem:

> *"...for the sake of the common survival, the comfort of personal property should be considered unimportant. Villages and farm buildings should be burned in this area in all directions, as far as the enemy [Romans] were likely to be able to seek forage and grain. The Gauls themselves had ample supplies of these kinds, because they would be supported by the resources of those in whose territory the war was being fought. The Romans, however, would either prove unable to bear the deprivations or have to go so far from their camp that they would incur great danger. And there was no difference between actually killing the Romans and depriving them of their animals – for when they lost these, they would not be able to continue the war. In addition to this, Vercingetorix said, all towns should be burned unless their fortifications or natural protections rendered them safe from every danger, for such places should neither serve as refuges for those Gauls shirked military service nor offer the Romans opportunities to seize rich supplies or plunder. If these measures seemed burdensome or harsh, then the Gauls ought to realize that it would be much more painful for their wives and children to be dragged off into slavery while they themselves were put to death: for this would be the certain fate of the conquered."*

Already the reader can apprehend the intent behind Vercingetorix's scorched-earth orders: (1) deny the Romans food for sustainment, (2) compel the Romans to make themselves vulnerable by causing small elements to become separated from the main host and therefore open to being defeated in detail, (3) demobilize the Romans by forcing them to eat their pack animals as soon as provisions start to become strained, (4) deny the Romans easy quarters for troops and cover from the elements, (5) ensure the greatest amount of mobilized Gauls to arms and to support the cause via denying the option of neutrality, (6) disincentivize a war against the Gauls by making it unprofitable. It is through the stratagem of scorched earth that Vercingetorix put the initiative back in his control and forced Caesar to act according to the circumstances the Gauls set, not the other way around.[77] Strategies should not just be confined to the battlefield but should be circumspect and address all the facets of military power, even those which are not readily self-evident. Shore up any advantages you have, deny any advantages to your enemy, and you will tip the balance of Fortune in your favor on the fields of Mars.

28

MAINTAINING CONFIDENCE IN YOUR COMMAND

"Be technically and tactically proficient."

— **USMC LEADERSHIP PRINCIPLES**

"Every soldier must know, before he goes into battle, how the little battle he is to fight fits into the larger picture."

— **GENERAL B.L. MONTGOMERY**

AMONG THE CHALLENGES of leadership, since any organization of man is indeed composed of men and all that entails, is the limits of self-discipline in the rank-and-file. Moreover, the position of the enlisted man, who is expected to obey orders, is often put in a position far removed from the nerve centers of command. This degree of separation from the sensory faculty of the host therefore leads to the situation where decisions and results are often not relayed which in their turn causes a type of morale dissonance to set in where one feels that the orders being received are not adequate or are given in the wrong "direction." Worse, if this state of dissonance is left to fester either by lack of communication or by incompetence on the part of the chain of command, the performance of the unit will suffer or lead to an all-out breakdown in discipline.[78] Vercingetorix himself is almost a victim of mutiny when the results of his Fabian stratagem were seen to be ineffective. Caesar recounts to use his circumstances here:

> *"When Vercingetorix returned to his men, he was accused of having betrayed them, because he had moved his camp nearer to the Romans, departed with all the cavalry, left such a large force without a commander, and offered the Romans such a great opportunity for attacking them as soon as he was gone. All this, they complained, could not have occurred by mere chance without his deliberate planning: obviously he preferred to have the kingship of Gaul by Caesar's permission rather than as a gift from themselves."*

To the charges of betrayal made against him, Vercingetorix replies thusly:

> *"As far as moving the camp was concerned, he had done it because of the shortage of forage – and they themselves had been pestering him to do so. As for going near to the Romans, he had been motivated by the suitability of the new camp's location, which recommended itself for being defensible without fortifications. The efforts of the cavalry would have made no difference in a swampy area but had proved useful in the place to which it had been sent. He had on purpose not handed off the*

> *high command to anyone when he left because he was concerned that the collective enthusiasm of the troops would force such a man into a decisive battle. He could see that weak-mindedness made all of them eager for just this kind of action, because they were incapable of enduring hardship any longer. ... He did not want to gain any supreme authority from Caesar through an act of betrayal when he could get it by a military victory, which already was assured for him and all the Gauls."*

In the dialogue, we are revealed two leadership principles from which his commander's mandate is restored. First, Vercingetorix explains to his men the reasoning behind his actions and the conveys the soundness of his tactical reasoning. Everything he did was with a purpose, and each tac he chose was a superior move to any other possible choice he could have made. The necessity of explicating himself was vital since it drove home to his men that he was in fact a competent leader, which in turn, shored up support for him continuing his command on that basis alone. Secondly, he explains to the ranks their own purpose in the scheme of maneuver, which in having acquiesced to such orders, from their view, seemed asinine beforehand. Vercingetorix explained to the men each of their individual purposes by default and therefore allowed them to conceptualize, with satisfaction, their own little part in the grander whole thus guaranteeing their continued support for him and their continued disciplined performance for their respective duties.

As a leader, you yourself must be an able communicator to your troops. It is true that a subordinate should always faithfully execute orders handed down to him, however, that is not a practical reality. Moreover, explaining to your subordinate leaders and down the chain of command is important because it allows for the individual teams to act better under pressure with more flexibility to face challenges because they are able to maneuver around any obstacles presented by direct orders and still accomplish the overarching mission. In short, it delegates responsibility to your troops, making them more invested and effective in their charge. Most important, in war you are expending the lives of men based on your judgement, it is right and just for men under the command of a leader to expect him to be competent and to not throw away their lives needlessly. In sum, legionary, make sure you know your craft, arrive at them decisively and with good judgement, and make sure to lead down your chain of command by keeping your men appraised.

29

THE ROAD TO RUIN IS PAVED BY SLIPS OF DISCIPLINE

During the succeeding campaign, Roman forces managed to encircle and invest the Gallic city of Avaricum. A significant settlement with equally formidable defenses, the Gauls had chosen to retain the town in a bid to store strategic supplies for the continuance of their campaign against Caesar. This, however, came to their disadvantage since it was the Romans who were superior in terms of siege craft expertise and the military technology to make the taking of the city all but inevitable. Early in the siege, however, the Gauls through their extraordinary heroism were able to fight the attacking Roman force into a stand-still where progress in the siege was being successfully frustrated. Over the weeks that transpired, Caesar noticed that the Gallic sentries slackened in discipline during times of rain and bad climate and it is here that he exploited the gap in the adversary's defense:

> *"The next day a siege tower was moved forward, and the siege works that Caesar had begun to construct were finished. At the same time, a heavy rain-storm had arose. When Caesar saw*

> *that the Gallic guards were deployed along the wall a little less carefully, he considered this to be a quite favorable moment for a well-planned move. He ordered his own troops to go about their work more sluggishly, and he explained to them what he wanted to happen. Unseen under the protective siege roofs, the legions got ready for an attack. Caesar urged them finally to reap the fruits of victory that all their hard work had earned them; he offered rewards to those who climbed the wall first, and then gave the soldiers the signal. They made a sudden swoop from all directions and quickly took control of the wall."*

The fortunes of war had swung decisively in the favor of the sons of Romulus. Owing to the continued discipline, soldierly proficiency, and fortitude through adverse conditions, the Romans were able to seize upon the opportunity given to them and snap the neck of the defending Gauls in one decisive action. The ensuing sack of the city was incredibly violent. In a fury for revenge for wronged fellow citizens who had been massacred in the same defenseless way, the legionaries tore through the city like a merciless red current killing men, women, children and the old. The disaster visited on the Gauls that day was so complete that only eighty of the city's thirty-thousand members survived the Roman vendetta. It is likely that this catastrophe could have been avoided had the Gallic commander insisted on disciplined sentries on the wall, and discipline in general, no matter the time or place. It was precisely this laxity which spelled the doom of all those who dwelled within the walls of Avaricum.

As warriors, and especially commanders, it is our duty to maintain discipline through thick and thin, because ultimately, as the Romans showed us in this dreadful episode, the enemy is always waiting for a moment of weakness to strike their deadly blow. Therefore, legionary, be disciplined and never let your guard down!

30

IMBROGLIO AT GERGOVIA: Caesar Disciplines His Over-Eager Legionaries

"Good initiative, bad judgement."

— USMC COUNSELING PROVERB

"From his soldiers he needed discipline and self-control as much as courage and greatness of spirit."

— CAESAR

WHEN IT COMES to the morale of one's men there can be two problems: on the one hand they can be cowardly and of lax willpower, or they can be impetuous and over-eager. Out of the problems which arise in the challenge to command, a leader wants his men to be of the latter rather than the former. Both, however, are caused by the lack of discipline, causing a circumstance where emotions rule over the hearts of soldiers and not the commands of their leader. Impetuousness of the men can lead to disaster just as easily as cowardice can. This is precisely the fate that befell Caesar in one of his few actions that he lost.

While maneuvering in the south of Gaul, Caesar was finally able to fix the Gallic forces at a place called Gergovia. This small town was strategically situated in such a place which bisected Roman communications and lines of supply between themselves and the local Gallic allies who were sustaining Caesar's operations in the region. It was during this time that, by a *ruse de guerre*,[79] the Gauls were able to turn Rome's most powerful client against their Roman masters. The fiasco with the Aedui, who were responsible for supplying and protecting these lines of supply to the Roman legions, induced Caesar to march out with half of his army to meet the oncoming Aeduian host. Due to Caesar's tact and gentle diplomacy, he was able to inform the Aedui that they had been tricked and co-opted the tribe back into supporting the Roman cause.[80] This positive development however came at the cost of leaving his force-in-readiness exposed behind him, which in turn came under brutal assault from Vertcingetorix's forces. The assault was successfully repelled by the time Caesar had returned but the casualties were too numerous to maintain their current position. It was in Caesar's interest, therefore, to retreat and link up with Labienus' four legions to the north in a bid to concentrate all the Roman forces to crush Vercingetorix once and for all.

The plan to retreat was sound, however, Caesar needed to make sure his strategic redeployment came from a position of perceived strength; otherwise, his Gallic allies would think the Romans weak and ultimately defect to Vercingetorix and thereby galvanize the rest of Gaul to rise to the occasion of common liberty from Rome. Therefore, Caesar put forward a daring plan which involved a quick raid into Gergovia before signaling the retreat.

The Roman battle plan were as follows:

SITUATION: The majority of the Gallic forces were camped defending a vital hill opposite the Roman fort to defend the passage to foraging the countryside, which therefore left the town wide open to assault.

MISSION: Caesar intended to exploit this gap in the Gallic defensive posture to raid into their headquarters, destroy as many provisions and key leaders as possible, score a psychological victory over the Gauls, and retreat in an orderly fashion.

END STATE: Having successfully executed the intended operation, it was Caesar's aim to demonstrate to both friend and foe alike that the Romans still very much maintained the upper hand and therefore any follow-on actions would be a play for more strength and not a projection of weakness.

The first points of the command's plan were successfully achieved during the ensuing course of the attack, however, it was in the succeeding events thereafter which imperiled the entire enterprise. Caesar recounts the decision point here:

> *"Having achieved the goal he had set for himself, Caesar ordered the signal for retreat to be sounded. He was with the 10th Legion and stopped its advance immediately. But the soldiers of the other legions did not hear the sound of the trumpet, because quite a wide ravine lay in between them and the trumpeter. Still, in accordance with Caesar's instructions, the military tribunes*

and legates tried to hold them back. But the soldiers were carried away by their hope for a swift victory, the flight of the enemy, and memories of past successful battles. They thought that there was nothing so difficult that their bravery would not allow them to accomplish it, and they did not halt their pursuit until they were close to the town wall and gates. Shouts then rose from every part of the town. Those who were some distance away were terrified by the sudden uproar; thinking the Romans were with the gates already, they burst out of the town....

When Caesar saw that the fighting was taking place on unfavorable terrain and that the enemy forces were increasing, he grew anxious for his men. He sent a messenger to the legate Titus Sextius, whom he had left to guard the smaller camp, and ordered him to bring the cohorts swiftly out of the camp and station them at the bottom of the hill on the enemies' right flank. That way, if Sextius saw our men dislodged from their position, he could frighten the enemy and prevent them from freely pursuing our men....

Our men were under pressure from all sides. They had lost forty-six centurions and were driven from their position. But the 10th Legion, which had positioned itself on slightly more favorable ground to support them, slowed down the Gauls when they pursued them too ruthlessly. In turn, the cohorts of the 13th Legion that were led from the smaller camp by the legate Titus Sextius and had occupied some higher ground covered the retreat of the 10th. As soon as the legions reached level ground, they stopped and turned their standards toward the enemy, ready for an attack. But Vercingetorix led his troops from the base of the hill back into his fortifications. Nearly seven hundred of our soldiers were lost that day."

In sum, the overzealousness of the men had the entire strike-force in peril had it not been for Caesar's expert commandership and foresight which stationed the crack 10th Legion along their egress route to cover their retreat and shield them from any possible pursuit by the enemy.

After the day had been concluded, the men had returned to the fort ragged and bleeding from fighting, returning to their quarters to take stock of the dead, wounded, and to repair any weapons lost in the fight. Next morning the legions were stood to attention and addressed by the commander himself. Here Caesar disciplines his men admonishing them:

> *"The next day Caesar called an assembly and took the soldiers to task for their recklessness and greed, chastising them for using their own judgement as to where to advance and what action to take, for not stopping when the trumpet call for retreat was blown, and for refusing to be held back by the military tribunes and the legates. He explained what impact unfavorable ground could have: he himself had kept this in mind at Avaricum, where he caught the enemy without general or cavalry but had forfeited a virtually certain victory in order to avoid even a small loss that might be caused by fighting on uneven terrain. As much as he admired the enormous courage his men, whom neither the camp's fortifications nor the hill's altitude nor the town's walls had been able to hold back, as much did he have to condemn their lack of discipline and, yes, arrogance – that they had thought they understood better than their commander how victory could be won and how everything would turn out. From his soldiers he needed discipline and self-control as much as courage and greatness of spirit."*

In counseling his troops, in the aftermath of such a near-disaster, Caesar exercises much self-restraint when giving constructive criticism. Note how he makes sure to mention the positive qualities that he wants from his men, their courage and intrepidity. It is only after that he really drops a heavy tone and dresses them down for their lack of obedience to orders. The tac he chose in disciplining his men is important: *to make your team better you need to make sure they know the positive things they have been doing under your charge before you gently advise them to improve on their explicated shortcomings*. What is all the more important is the measured tone Caesar took with his troops, despite having, by cause of the legions indiscipline, almost lost the lion's share of his command the day prior which would have therefore severely imperiled the entire campaign in Gaul!

Fatherly leadership is a foundation stone for commanders through time, and it is readily apparent here in the tac Caesar chose to correct his troops. Make sure you laud your troops for their successes and only then follow with tactful, but firm, admonishments when necessary. Most of all remember that, as a leader, *everything your team does or fails to do is your responsibility*. Just as their successes are your successes, so too are their failures. When you take ownership of their faults and virtues you become invested in their success. When your men feel your investment and fatherly tough love, I promise you, they will rise to the occasion just as the sons of Caesar did all those centuries ago. May Mars bless you with such sons so that you may forge your own Rome!

31

PREEMPTING THE ENEMY

"An ambush, if discovered and promptly surrounded, will repay the intended mischief with interest."

— VEGETIUS

IT WAS IN the following period of maneuvers where the Roman forces were concentrating and escorting Roman citizens to the safety of Province that Vercingetorix decided to strike. The Gallic commander was quick to pick up that he held a distinct cavalry advantage over his foe, a superiority in morale, and most importantly, the Romans were hindered by civilian contingents that they were escorting. It was the intent of the Gauls to exploit the Roman's weaknesses at this most opportune time by laying an ambush for the Roman host before any more reinforcements could leaven the numbers under Caesar's command. Caesar, for his part, preempted this move and had already put out a call for allies in Germania to come to his aid by promising payment and booty, to fight for him and act as his reinforced cavalry contingent. Caesar went so far as to make sure these mercenaries were provisioned with adequate mounts since the steeds they had ridden in on were of inferior quality. In an otherwise operationally weak position, Caesar was able to leverage his critical vulnerabilities in his favor to lure the Gallic enemy into a trap which he could then use to score a decisive victory in furtherance of his campaign.

Caesar, preempting his enemy and understanding his decision-making cycle allowed him to seize back the initiative in Gaul. Caesar tells us of the action when the Gauls made their attack:

> "*This course of action was approved, and the oath was administered to everyone. The next day the cavalry was divided into three contingents, two of which were displayed on either flank of the Roman column while the third proceeded to hinder the march in the front. When this was reported to Caesar, he ordered his own cavalry as well to split into three divisions and move against the enemy. All at once there was fighting on all sides. The marching column halted and placed the baggage in the center of a square formed by the legions. If anywhere our men appeared to be in trouble or especially hard-pressed, Caesar ordered the infantry to move forward in that direction and take a stand in battle formation. This slowed down the enemy in their pursuit and encouraged our troops by giving them the assurance of support.*
>
> *At last, the Germans on the right flank reached the top of a ridge and drove the enemy from it. They chased the routed cavalry to the river, beside which Vercingetorix had taken a position with his infantry forces, and killed a large number of them. When the rest of the attacking Gauls noticed this, they were afraid of being surrounded and sought salvation in flight. There was slaughter everywhere... With his entire cavalry put to flight, Vercingetorix*

> *withdrew his troops from where he had stationed them in front of the camps and immediately began to march towards Alesia."*

In the exchange above it should stick out to any commander the tactical proficiency up and down the Roman chain of command. However, Caesar's mastery of the tactical and technical is beside the point, since it is the added element that Caesar brought to the battlefield which snatched victory from the jaws of otherwise certain defeat. Before the battle was met, and the maneuver was planned, Caesar had anticipated the Gauls' wish to exploit their successes by striking a Roman opponent when they were down.[81] Therefore, he wanted to leverage the ambuscade that surely was going to be waiting for the Romans in the days ahead by raising further forces which would act as flankers to an enemy which had intended to take their targets by their own flanking maneuver.[82] It is this move which showed the genius of Caesar and is a principle you, legionary, should remember best. Make sure to be thinking from the perspective of your enemy and use his forecasted decisions against him. Constantly be making provisions for contingencies, and once committed to battle, follow through with ruthless determination!

32

DECISION AT ALESIA

"The strong do what they can and the weak suffer what they must."

— **THUCYDIDES**

IN THE PROBABILISTIC philosophy of war[83] it is a necessary element that we seek decisive battle whenever possible and wherever the success of which may, weighed against the risks, lead to an operational or even strategic victory. By nature of seeking decisive engagements, it is possible that in that bid for victory you are forced to sup from that bitter chalice called defeat. The balance between defeat and victory in such an occasion comes down to three decisive elements: discipline, fortune, and willpower. All three of these were leveraged to the hilt at Caesar's greatest victory at Alesia.

When Vercingetorix arrived at the stronghold of Alesia he was still being actively pursued from his last battle with the Romans. German auxiliaries and Roman cavalrymen were nipping at what remained of the Gallic host who were still trying to find refuge within those fateful walls. The slaughter in this retreat was so immense that it effectively cut the Gallic force by a further quarter on top of the original casualties incurred from the battle preceding it. Vercingetorix, however, being the great leader that he was, managed to buoy morale once more and steeled the nerves of his compatriots to prepare for a siege. Thinking ahead, just before the Roman circumvallation had been completed, he sent out messengers and families of those from the wider Gallic nation to head back to their tribes and call for aid. His beleaguered request was met with unparalleled patriotic fervor as Caesar describes here:

> *"There was in all of Gaul such a powerful and unanimous desire to restore their liberty and recover their old-time martial glory that people were moved neither by favors they had received nor by the memory of friendship. Instead they threw themselves into this war with all their passion and resources. Eight thousand horsemen and around 250,000 infantry were assembled; in the territory of the Aedui they were all marched out for review, a tally was made, and officers were appointed. The high command was given jointly to Comminus of the Atrebates, Viridomarus and Eporedorix of the Aedui, and Vercassivellaunus of the Arverni, Vercingetorix's cousin. Men chosen from the various nations were assigned to them as advisors on the conduct of war. Everyone was elated and full of confidence as they departed for Alesia. Not a single man among them all doubted that the mere sight of such an enormous force would overwhelm any resistance, especially in a battle on two fronts, when one army would fight in a sortie from the town while on the outside such huge numbers of cavalry and infantry were being displayed."*

Even before Caesar began hearing reports of Gallic mobilization, he had presaged a strong reaction and therefore set upon creating a contravallation[84] to protect the siege lines from being penetrated and thus letting Vercingetorix escape his grasp. The siegeworks built around Alesia were extensive with three sets of barriers[85] before the rampart was reached, which in itself formed the principle defensive line. As you can imagine this drudgery was exhausting, your average legionary often complained just as your modern grunt often does, however, it was this iron discipline of the Roman legionary that helped secure their victory ahead of time. Over ten miles of distance were encrusted with extensive siegeworks bristling with stakes, moats, and other redoubts. It was only after thirty days of grueling siege that the relief Gallic host arrived, a quarter of a million strong, blaring their deep horns of battle and striking the war drums deep into the night.

As soon as the battle was met combat was grueling and fierce. Repeated assaults on the Roman positions were carried out but none proved to be decisive nor particularly lasting. Every time a sally had been made it was successfully beaten back and the ramparts were reconstructed. Finally one sally culminated the days of combat, and it is here where Caesar snatched victory, riding into battle adorned with his purple cloak of command, steeling the spine of his troops against the Gallic foe, and etching his name into the scrolls of history:

> *"The color of the cloak that Caesar habitually wore in battle to mark him out as commander made his arrival known to the enemy. They also spotted the cavalry squadrons and cohorts Caesar had ordered to follow him, since the lower slopes and depressions they passed through were visible from the higher ground. Their appearance prompted the enemy to renew their efforts in the battle.*
>
> *Both sides raised a shout, and those on the rampart and on all the fortifications took it up in their turn. Our men no longer relied on their throwing spears but got busy with their swords. Suddenly the enemy became aware of the cavalry approaching at their back. The cohorts just mentioned were advancing on*

> *them. The enemy turned to flee. The cavalry ran into those fleeing. A great slaughter ensued. Sedullus, commander and leader of the Lemovices, was killed. Vercassivelaunus the Arvernian was capture alive in flight. Seventy-four military standards were brought to Caesar. Out of that large number, only a few escaped unhurt into their camps."*

Shortly after that sharp and decisive defeat, Vercingetorix decided to capitulate, hoping that Caesar would be merciful and enslave, rather than butcher, the remaining Gauls within Alesia. Caesar made a point to note in the Commentarii de Bello Gallico to note Vercingetorix's selflessness and courage to offer himself as a living captive in hopes to spare his people.[86] The Romans accepted Vercingetorix's capitulation, had him ride into the camp of the Romans behind the tall walls they had constructed, descend his steed, throw down his arms, prostrate before Caesar enthroned, and give the kiss of peace to the Eagle of the Legions. The inhabitants of Alesia were enslaved and the spoils within were split among the ravenous legionaries. Roman victory was total, drawing to a close the siege of Alesia, the rebellion against Rome, and the wars in Gaul as a whole.

Militarily speaking, there was no reason why Vercingetorix should have lost that action at Alesia. The Gallic forces were far stronger and just as capable as their Roman counterparts, if not as organized or well disciplined. The Gauls could have overwhelmed the enemy had they attacked along more points in the Roman defensive perimeter thereby overwhelming the many times outnumbered defenders.[87] The result of having taken the above mentioned tac would have likely led to the annihilation of the Roman army as a force-in-being and likely have secured a decisive independence for a united Gallic nation under Vercingetorix as a great king. All this was not to pass, however, and tactical mistakes like this are often the very thing which can gain or lose a war. In battles of decision, a single stroke of a sword, a minute more of austere discipline, a single roar of encouragement at the moment of truth is just what may be needed to win. And all of this is trained, for years at a time before this veritable moment of truth.

Therefore, legionary, understand your professional calling. Take ownership in being tactically and technically proficient in your job. Cultivate an indomitable will to power, and I promise you, that preparation will be the weight of fortune in your soul that steadies your hand and readies your heart to overcome any challenge. Excellence is a habit trained every day of your life. Remember that, and the world is yours just as Gaul became Caesar's that day on a muddy battlefield around a small town which would be forever famous.

CONCLUSION

EMBLAZONED IN THE diadem of the Western military tradition there exist many great and awe-inspiring names. Among those etched into the annals of military greatness are names such as Alexander the Great, Napoleon, Scipio Africanus, Fabius, Pericles, Achilles, Richard II, Rommel, Agamemnon, Patton, Stonewall, and Marshal Jomini. Yet, it is only one name that inspires a glimmer in the eyes of many a young legionary aspiring to both martial and civic glory. Julius Caesar was a man of near forty years of age before he strapped on his caligae in earnest.[88] Many believe that old age is a bar to military prowess or activity as such, yet Caesar serves us a shining example of what one can do if one has the will, regardless of circumstances or personal inhibitions. Moreover, Caesar serves us as a model of what we, professional warriors, should strive to emulate, in general, but specifically in the time between war.

Caesar was a consummate student of history, philosophy, and the military arts, even when the mechanisms of acting out on those ideas were not present in his life. Caesar's purpose in studying these men, through their recorded words, was to arm himself with the experiences of others during his leisure time. The Roman nobility imbibed a concept whose motto went "Otium et Bellum"[89] which is in English, "Leisure and War", two states which every Roman aristocrat sought to spend his life in. Every aristocrat is by nature a warrior, he is a man who set upon himself the duties and disciplines of military training, whether through times of peace or war, in which the retention of such skills are indicative of not just his military acumen but serves as a foundational element which augments his overall excellence[90] and *Romanitas*.[91] It is with this ethos that I wish for you, legionary, to imbibe in your own life, regardless of your military service or station in life. What delineates a civilian from a warrior has less to do with what others have socially recognized you to be and has more to do with what personal bearing, training, and moral convictions you exude. Therefore, be akin to Caesar, whether you are young military officer just starting his career, a veteran who has been apart from the service for decades, or if you are a citizen-soldier looking to become what you never had the chance to be, the prescription is the same: study war, its art and sciences, but most of all, study the campaigns of great men who have preceded you and analyze their commands for your own edification.

Caesar Bellator started as an extension of my personal interest in Campaign Studies which I used to teach my peers and subordinates the tactical and operational elements of war. The exercise for most is informative enough regarding the history of events, however, simple chronology is not the aim nor are the sets of analyses I have posited prescriptive. Instead, the intent behind any given Campaign Study is to place yourself in a state of mind which forces you to *make decisions*. In other words, you shouldn't be reading military history and remembering all the different solutions to unique problem, instead, you should be searching for perennial principles which are applicable regardless of unique factors and which can be adapted to your uses outside of any ideal circumstances. Lastly, use this principle to train your subordinates and peers so that they can start to understand the principles of honing the military art.

Legionaries, it is my great pleasure that you have read this small book and taken it to heart. However, the path to the mastery of war does not end here, it is only a beginning. Hone your art, your science, and your body to the discipline everyday. Give yourself totally to the Mission assigned to you, follow it down whatever path it may lead you, even if that path leads to death. Live this life by the core principles of Honor, Courage, and the Commitment to

Excellence and even the celestial kingdoms cannot resist your onslaught! Give honor to the god of war, Mars, and he will champion your cause!

MARS EXULTE!

ENDNOTES

1 Leaving The Gallic Wars and other documents for both contemporary readers and posterity.

2 Both figuratively and literally since he was bestowed divine status upon becoming Dictator of Rome.

3 "I love the name of honor more than I fear death." — Julius Cesar

4 This circumstance should sound extremely familiar to you if you live in the West.

5 Ironically motivated to restore the Republic as it was, and I believe it was done so genuinely, he was a key factor in normalizing a Dictator archetype which secured power through force and not through legal means.

6 This is called the "First Triumviratte."

7 This is an old Roman custom where Consuls would be given a province to squeeze wealth from through corruption and extortion.

8 Caesar was governor of three provinces simultaneously: Cisalpine Gaul, Trisalpine Gaul, and Illyricum. Two elements of this term in office were unprecedented since no one before him held multiple provinces simultaneously nor did they maintain it for longer than a year.

9 In fact, there is an old Roman legend about the first sack of Rome where the Gauls successful overwhelm the defenses of the city and all

those who were left were the old men in armor who died giving it their last. It was both an emotional and beautiful story which buttressed Roman spirits during the Punic Wars and beyond.

10 More often than not this assimilation came in the customary form of slavery.

11 Caesar in fact names them the second most manly tribe of the Gauls where he instead names the Belgae as the most violent and fearsome.

12 The British Empire even commissioned a book which catalogued the martial peoples which were under their reign. Among them were the Nepalese who inhabited a very similar terrain to the ancient Helvetii.

13 Often in the form of a state, Sparta is a famous example where Lykourgos restructured and regimented his warrior society and artificially created an environment where wealth and luxury were forcibly prohibited. Famous among these innovations were his new issuance of iron coins instead of gold, of equal sized lot housing, and the communal mess halls that all men were required to take sustenance together regardless of rank or status.

14 Kingship was not common in the commune-like tribal living habits of the Gauls.

15 This concern was also militarily risky because the inhabitants of this territory were hostile to the Helvetii and if they chose to defend their pass it would be very easy to block any advance due to the favorable terrain. Caesar mentions that this terrain is so difficult a bulwark that could be defended successfully with only a handful of men.

16 The bridge at Geneva was the only place over the Rhone for an army to cross for roughly two-hundred miles. Deny the enemy passage allowed the Romans to deny them movement or risk canalizing themselves in an assault.

17 After there was a negotiated settlement and a customary exchange of prisoners for the Helvetii to transit through their territory peacefully.

18 Sustenance is the most important aspect of logistics in this time. Human beings were often stunted in growth due to a lack of adequate nutrition. The access to a good source of food had physical ramifications for nations and their warfighting potential. Remember that Napoleon said, "an army marches on its stomach."

19 A great example of "Soft Power."

20 This is an essential element, when dealing with hearsay and to guard oneself against disinformation or subversion, in these circumstances of measuring the veracity of hear-say it is important to dot your I's and cross your T's. A false accusation can lead to dissent in your political support base and ultimately harm yourself to the Enemy's benefit.

21 The American effort in Iraq and Afghanistan fell into this trap where they were unable to secure the loyalties or sympathies of the populace because of the ham-handed self-righteous approach to trying to "nation build" which in no uncertain terms means Paraphilic/Communistic social programs which alienate the local populace (as well as the vast majority of American citizens back home). Every failure against "insurgencies" in America's past has come down to this inability to be sensitive to the wishes and foundational values of the people they are attempting to subjugate.

22 Again, the de-Ba'athification of Iraq was a major factor in the continued insurgency and instability in the Iraqi state. The Americans basically deprived men of the only employment they knew (government/military) and thus were unable to make a living. Backed into a corner and increasingly more angry and entrenched in their resentment they began a meaningful insurgency. To pacify a populace, you have to offer a way out that is a viable way

to live and it has to be more appealing as an alternative (a.k.a. you need carrots).

23 Ariovistus was a man known to be particularly fearsome and bloodthirsty. He was quick to anger to irrational heights and even the most diplomatic approaches were received with his violence.

24 "Suebi" was used as a term among the Romans as a generic moniker for Germans in general.

25 Vesontio was a strategic location because it was the only major center for materials of everything useful in war from arms to food and fodder. Additionally, it had the significant advantage of being surrounded on all three sides by a flowing river and being sited on a hill making it a significant defensive position for anyone who is able to occupy it.

26 Feigning difficulty or illness to extricate oneself from duty.

27 Fear is contagious, a leader has the duty to be a pillar of certainty and resolution among his men for them to be continually reassured and confident in themselves.

28 This has the effect of undermining the confidence in their "un-confidence" while mentally framing the coming message in a positive context.

29 This again has the effect of mentally framing the coming challenge as smaller than the enemies they have already bested.

30 This had the double effect of buttressing the self-confidence of the Tenth and challenging the other legions out of shame to be shown up.

31 It is important to emphasize that the kind of shame a leader uses should never be too strong or come off as a "scolding." You want to make sure there is an underlying tone of genuine respect for your

men and confidence in them. It should be nothing more than a nudge or you risk being despised and for the men to become combat ineffective due to crippling the Will via self-doubt.

32 It is important to remember that among the duties of a Proconsul, aside from the military command and responsibilities, was to hear and adjudicate petitions and make judicial rulings which had accumulated over the campaigning season. Caesar, of course, had innumerable scribes, clerks, and skilled slaves to help with the work load of governing an entire province, however, you must understand the immense strain this continued to be on any individual. Caesar, who was both a master of both military and governing affairs, also found time to write his commentary on his various wars. The titanic energy that Caesar had was foundational to the success of his campaigns as well as for his political exploits.

33 When looking back in history, it is important to note that things are not predestined, nor is it necessary that one event should have played out the way it did. Success, as Machiavelli puts it, is half Virtu (skill and spirit) and have Fortuna (luck). Just as your life is now, events and historical occurrences are fluid and can fall one way or another.

34 "Life itself is essentially appropriation, injury, overpowering of the strange and weaker, suppression, severity, imposition of one's own forms, incorporation and, at the least and mildest, exploitation..... because life is Will to Power"— Nietzsche

35 Contentment is a sign of abated and dying Will and life force.

36 In the form of spies, diplomatic missions, and correspondence with Roman garrison commanders.

37 Historically speaking, armies until the industrial age were much limited to operations within a timeframe where the temperature was favorable to movement and foraging. Generally, this was between the months of March and October. March, by the way, is derived

from the Roman God "Mars" due to it being the historical month that Roman armies would first be able to muster the levy i.e. "Legio" before the Marian reforms.

38 It was customary for subjugated nations to offer provisions of food and weapons in addition to whatever detachment of warriors they were able to produce as auxiliary forces. The most important function they served was as of guides and translators.

39 Defeating an enemy in detail is perhaps one of the most overlooked but essential elements to defeating any enemy. Napoleon, who was faced with similar force disparity in his Italian campaign, employed this Operational tactic so that he would have local numerical superiority and therefore superior combat power compared to the Austrians, even though the Austrian Empire had deployed a far larger army in that theater.

40 Its important to note that in classical warfare, the majority of casualties were accrued during times of routes. No quarter was given and death was the order of the day.

41 Although it is important to mention that this clemency came after taking a number of hostages to ensure their continued good faith.

42 This happened because the progress of the baggage train would hinder the swift arrival of the second half the legions due to the fact that the surrounding wilderness was impossible to be traversed effectively unless by road.

43 In fact, when the attack was met many legionaries didn't even have time to uncover their crests or shields. Nor did they have time to down their packs properly. Such was the total surprise that the Roman army was held in.

44 Once a friendly unit falls into a route, it is likely that adjacent units succumb to the panic as well. Though the domino effect is not born

out of cowardice but is because those adjacent units which relied on the brothers that were fleeing made their position untenable.

45 Their location was positioned in the modern Brittany region, not to be confused with modern Venetia.

46 Caesar insinuates here that they were tortured and possibly sexually predated upon.

47 Classically known as "breaking a people."

48 Scholars in modern Communist academia have taken the stance that this was the habit of "conservative" Romans trying to criticize the Rome of their time by juxtaposing themselves against a people who emulate Rome's founders and the way things ought to be. It's important to annihilate this paradigm from your mind because here is the thing: Academics, by their very nature, are weak-bodied, weak-willed, and petty individuals. Scribblers like them parse any information they are presented with through the lens of weakness and therefore obscure a greater truth which requires strong hearts to apprehend. Men like Caesar, Tacitus, Suetonius and so on were all military men who knew how to admire an adversary for their virtues without being betraying themselves to them. In short, when these men tell you about the virtues or vices of a foreign people I believe it is prudent to take that statement prima facie since strong men are always forthright in telling the naked truth.

49 There is a good anthropological case to be made that the Dorian Greeks and the Teutons were more closely related to each other in the time of Caesar than the Dorians were to the Ionians.

50 "Lycurgus the lawgiver, in his wish to convert the citizens from their existing habits to a more disciplined way of life and to make them brave and honorable (since they were living a soft life), reared two puppies born of the same father and mother; and one he conditioned to a life of luxury, allowing it to stay at home, while the other he took out and taught to hunt. Next he brought them into

the assembly, put down some bones and delicious tidbits, and then released a hare. Each of the two dogs went after what it was used to; when the second of them had caught and killed the hare, Lycurgus said: 'Citizens, do you see how, although these dogs belong to the same family, their upbringing for life has made them turn out very different indeed from each other?" — Plutarch, On Sparta

51 "Friend of Rome" was an official title in recognition for a foreign national that had done the Republic a good turn or other such service that earned Rome's good esteem and official support.

52 This is, in military lingo, referred to as "Mission Creep" where the original goals of the campaign or task are met and new ones are tacked on. This is a massive mistake and causes any campaign to founder as goals become increasingly ambiguous and the parameters of victory become nebulous. The United States has been notorious for doing this and is the reason why we have lost every major war since 1945 with the exception of the First Persian Gulf War.

53 Every time a Roman legion makes camp they always set up the usual square formation fort with the same internal layout so that any legion will be able to navigate any fort with the same dexterity and presence of mind at all times.

54 Especially when taking into account the exotic bio-mass which is imported into Britain by the Communist subversives of our time, although these will be expelled promptly when the Patriots take back their country and their future.

55 The Battle of Gaugamela (331 B.C.) was the last major engagement in the civilized world that saw the mass employment of this technology in battle.

56 Most unbecoming of any officer in any time or military. Officers are supposed to remain calm and give prudent and cynical realistic advice.

57 These were what was called "pila muralia" which were stakes that were used to push down enemies as they climbed the walls. These weapons could be thrown or used as an ordinary spear.

58 "Inspect what you Expect." — USMC Proverb

59 There is this concept about leadership called "the mask of command" which is almost a stoic self-portrayal of ones command self to others as almost an actor would. Now, don't get it twisted, this doesn't mean to be a charlatan. To have a good command presence it is necessary to act in a stoic, stolid, disciplined, and determined manner because most men look to their commanders as the weather vain for how things are going. One most constantly portray confidence and resilience to instill that in one's command. "Have discipline, give discipline" is a great way to encapsulate the reciprocal relationship there is between cultivating reality and perception. Remember, gentlemen, the best "mask" is one genuinely born. You exude what you are, your men will know if you are bullshitting.

60 Caesar (5.41)

61 So much of modern military leadership discourse revolves around "getting my men back home" which should only be a secondary consideration. The Mission comes first since it is the purpose of warriors to fight, kill, and if necessary, die. Its always important to remember that all men die, few have the privilege of dying fulfilling a grand purpose leaving behind a beautiful body and great deeds. Only women die in hospital beds at a disgusting old age, leaving behind a decrepit body. As the saying goes, "A ship is safe in harbor, but that's not what ships are built for." Care for your men as best you can, but always fix your mission unshakably in the furnace of your heart.

62 Arete (Ancient Greek: ἀρετή, romanized: aretḗ) is a concept in ancient Greek thought that, in its most basic sense, refers to

excellence of any kind—especially a person or thing's full realization of potential or inherent function.

63 As noted before, Cicero had sent riders when the enemy had originally invested but they had all been intercepted.

64 This includes judging the terrain, the quality of the enemy's morale and training, and that capabilities and limitations of your own men.

65 Here I am alluding to Lycurgus' holy suicide after having created the foundation of the Hellenes most prized people, the Spartans, as retold to us by Plutarch.

66 It bears repeating that Prussia held the lead for technological innovation, philosophical prowess, and political power for many centuries. The Soviet Union and the United States went to space on the merits of Prussian scientists.

67 *Galula, David, and John A. Nagl. Counterinsurgency Warfare: Theory and Practice. Pentagon Press,* 2010. P.2

68 Ibid. P.4

69 Though I will not explicate in detail, there is a story of a Centurion, Publius Sextius Baculus, who rose to the occasion and steadied flagging morale of the soldiers under attack through his own exemplary bravery. This is the stuff of military glory and success.

70 Think to yourself of the pain of labor of a mother who, after having suffered the pain and risk the possible misfortune of death, gives birth with no readily apparent benefit to herself becomes enamored to her offspring by mere virtue of the pain born to her. Or think of Scaevola, or any great patriot, whose every suffering impels him to even greater degrees of devotion to his Patria. The examples are innumerable, but the truth is constant.

71 "The lion cannot protect himself from traps, and the fox cannot defend himself from wolves. One must therefore be a fox to recognize traps, and a lion to frighten wolves."— Machiavelli, The Prince

72 The Gauls had assumed the mountain roads would be impassible under the still present winter snows which would keep the Roman legions at bay until the spring melts.

73 Among the militaries of our time which are the most notorious offenders of this key military principle is the United States military and that of the Prussian, then German, militaries between 1870-1945.

74 General Andrew "Stonewall" Jackson comes to mind and his Valley Campaign pushing his men beyond the limits of what was thought possible to attain decisive tactical and operational victories in the lead up to Second Bull Run.

75 Probabilistic View sees war as "seen as a complex phenomenon in which participants interact with one another and respond and adapt to their environment. The probabilistic viewpoint sees combat as unpredictable." (MCDP 1-3 Tactics)

76 Arete (Ancient Greek: ἀρετή, romanized: aretḗ) is a concept in ancient Greek thought that, in its most basic sense, refers to "excellence" of any kind—especially a person or thing's "full realization of potential or inherent function." The term may also refer to excellence in "moral virtue."

77 Therefore it put Roman forces on to a set of dilemmas which made them lose in one respect or another no matter how they chose and made their follow-on choices predictable to the Gallic command. This is in addition to the fact that time was now on the side of the Gauls which put further operational difficulties on the Romans on the way to trying to achieve a decisive victory.

78 Here I allude to mutiny, desertion, insubordination, or any other type of derelictions of duty in their varying degrees.

79 The French ruse de guerre, sometimes literally translated as ruse of war, is a non-uniform term; generally, what is understood by "ruse of war" can be separated into two groups. The first classifies the phrase purely as an act of military deception against one's opponent; the second emphasizes acts against one's opponent by creative, clever, unorthodox means, sometimes involving force multipliers or superior knowledge.
(Matuszczyk, A. (2012). *Creative Stratagems: Creative and Systems Thinking in Handling Social Conflict*. Kibworth/GB: Modern Society Publishing. (p. 21))

80 This episode in the campaign is overlooked here since the incident would be beyond the scope of the lesson. Suffice it to say that one should read the text and listen to how Caesar readily forgave the duped Aedui and tactfully allowed some major slights that happened among the Roman community in Gault to go unpunished for the sake of political unity and to win the war against Vercingetorix. The ability for a commander to overlook petty slights and focus on the grand scheme of things is invaluable and among one of the very important qualities any commander should have to leverage in the fight for victory over his enemies.

81 The move by Vercingetorix is sound and is a key principle of success in war. You don't incur the majority of your damage on the enemy by the engagement, but when they are being routed. Following up with a high operational tempo to exploit initial victories is vital.

82 Ambushing is strongly in the favor of the ambusher, unless of course, the ambusher is pre-empted and outflanked by supporting forces which would pounce on their rear once the ambushers initiated their attack.

83 MCDP 1-3 "Tactics"

84 A Contravallation is a second set of siege lines facing away from the invested site to defend against flanking maneuvers of

85 These barriers were formed by hidden pits with stakes, moats, and exposed siege-pikes fixed in place among other whicker works designed to slow the enemy's sally to make him open to enfilade missile fire.

86 7.81

87 The Germans would make a similar mistake at the Battle of the Bulge roughly 1,900 years in the future where instead of attacking along all sides at once they would attack in piecemeal along only certain sections of the surrounded perimeter. When fighting a surrounded force, make sure you are overwhelming the enemy along every front all at once and in force.

88 Though Caesar had some experience with the military before setting off to be Proconsul of Cisalpine-Gaul, it was largely negligible and therefore impossible to claim that he had mastered the martial skills requisite to make his conquest of Gaul manifest.

89 "The Roman aristocracy, as Nietzsche says, had the motto otium et bellum, leisure and war, these being the only right ways of life for a man of power and freedom."— Bronze Age Pervert, Bronze Age Mindset

90 Greek concept of Arete which I am alluding to once again.

91 Romanitas is a Latin word meaning "Roman-ness" or "Roman-hood." It is a concept that refers to the collection of political and cultural practices by which the Romans defined themselves. It encompasses the ideals and values that the Roman civilization held dear, such as citizenship, military prowess, and the idea of a unified Roman identity.

Printed in Great Britain
by Amazon

40386277R00098